Drugs That Changed the World

Drugs That Changed the World

How Therapeutic Agents Shaped Our Lives

Irwin W. Sherman

CRC Press
Taylor & Francis Group
Boca Raton London New York

CRC Press is an imprint of the
Taylor & Francis Group, an **informa** business

CRC Press
Taylor & Francis Group
6000 Broken Sound Parkway NW, Suite 300
Boca Raton, FL 33487-2742

© 2017 by Taylor & Francis Group, LLC
CRC Press is an imprint of Taylor & Francis Group, an Informa business

No claim to original U.S. Government works

Printed on acid-free paper
Version Date: 20160922

International Standard Book Number-13: 978-1-4987-9649-1 (Hardback)

Library of Congress Cataloging-in-Publication Data

Names: Sherman, Irwin W., author.
Title: Drugs that changed the world : how therapeutic agents shaped our lives / Irwin W. Sherman.
Description: Boca Raton : Taylor & Francis, 2017. | Includes bibliographical references.
Identifiers: LCCN 2016022679 | ISBN 9781498796491 (hardback : alk. paper)
Subjects: | MESH: Pharmaceutical Preparations--history | Drug Therapy--history | Drug Discovery--history | Social Change--history
Classification: LCC RM300 | NLM QV 11.1 | DDC 615.1--dc23
LC record available at https://lccn.loc.gov/2016022679

Visit the Taylor & Francis Web site at
http://www.taylorandfrancis.com

and the CRC Press Web site at
http://www.crcpress.com

Printed and bound in the United States of America by Publishers Graphics, LLC on sustainably sourced paper.

Contents

List of Figures

Preface

The dictionary defines *drug* as a therapeutic agent. Through the ages, drugs—natural or synthetic—have been used in the diagnosis, alleviation, treatment, prevention, or cure of disease. These drugs are based on thousands of years of knowledge accumulated from folklore, serendipity, and scientific discovery. This is a book about drugs that have changed the world in which we live, how they were developed, and the manner in which they exert their "magic." In telling the fascinating history of the findings and use of drugs, there is drama of triumphs and failures filled with colorful personalities—the selfless and the selfish, the competitor and the cooperator, the hero and the villain, those with unbridled ambition, those seeking fortune, and the bold as well as the timid.

Today, there are drugs to protect against infectious diseases, to alleviate aches and pains, to allow new organs to replace old ones, and to modify brain function. Yet, for the most part, the manner by which drugs are developed and by whom remains a mystery. For many of us, a drug is a pill or a liquid, prescribed by a physician, found in the medicine cabinet. Drugs are more than this, and although only several dozen or so drugs have markedly affected our lives and altered the path of civilization, to simply catalog these would be too numbing to read. Instead, I have selected a sampling—drugs that represent milestones in affecting our well-being and that have influenced social change.

Quinine, the first antimalarial to cure the disease malaria (Chapter 1), illustrates how a natural product enabled armies to triumph in the tropics. This chapter also deals with synthetic antimalarials (atabrine, chloroquine, and mefloquine) and the newest natural product, artemisinin. It provides the reader with an appreciation of what goes into drug testing, the role of animal models, the need for clinical trials, and the problems associated with drug resistance. In the second two chapters, Chapter 2—on aspirin, the first drug to treat simple pain; and Chapter 3—on anesthesia, the power of making a person insensible to surgical operation, are described. Hormones, too, are drugs; their role is illustrated in "The Pill" (Chapter 4), where a synthetic hormone analog revolutionized contraceptive choices for women; and in Chapter 5, where the isolation of insulin, the first hormone therapy for treating diabetes, is examined. Smallpox (Chapter 6) led to a vaccine that ultimately eradicated the disease and was the impetus for the development of other vaccines to combat infectious diseases (Chapter 7). Syphilis (Chapter 8) provided the spark for a cure through chemotherapy (salvarsan), which eventually led to the first antibiotic—penicillin, as well as synthetic antimetabolites (prontosil and pyrimethamine) for treating bacterial diseases (Chapter 9). The modern plague, AIDS, and the development of antiretroviral drugs are covered in Chapter 10. Organ transplantation became possible with the first drug, cyclosporine, to reversibly blunt the immune system (Chapter 11); and in the last chapter (Chapter 12), the relationship between malaria, madness, and the first psychotropic drug, chlorpromazine, is explored.

This is a book for the curious. It is for those who want to know more about the science behind the label on the prescription bottle or in the vial of vaccine. I have written it for those who wonder where the most important medicines came from and the societal consequences of their introduction. For me, learning about how drugs have shaped our lives and our civilization has been an exciting and enlightening experience. Hopefully, the readers of this book will also find this to be true.

Irwin W. Sherman
University of California at San Diego

A Note to the Reader

Interspersed throughout the text, chemical structures appear. These should not be a deterrent should your knowledge of chemistry be limited. Rather, the structures are provided so that those with the appropriate background and interest can appreciate the relationships between the chemical structures. Even without reference to these chemical structures, the text should be entirely comprehensible to all.

Parts of Chapters 1 and 9, from the author's previous book *Magic Bullets to Conquer Malaria* (Copyright 2011, American Society for Microbiology) are used with permission. No further reproduction or distribution is permitted without prior written permission of the American Society for Microbiology. In Chapters 6 through 8 and 10, material previously published in *The Power of Plagues* (Copyright 2006, American Society for Microbiology) is used with permission. No further reproduction or distribution is permitted without prior written permission of the American Society for Microbiology.

Cinchona calisaya. (From Köhler's Medizinal Pflanzen. With permission.)

Chapter 1

Malaria and Antimalarials

Malaria and the Earliest Antimalarial Drug

The disease malaria has had a long association with humans and historically assumed a variety of guises, being called intermittent fever, marsh fever, or simply "the fever." The Greek physician Hippocrates (460–370 BC) recognized that at the harvest time (late summer and autumn) when Sirius, the Dog Star, was dominant in the night sky, fever and misery would soon follow. Hippocrates believed these periodic fevers were brought about by drinking water drawn from the stagnant marshes. Malaria spread across the European continent, and in England, seasonal fevers called agues (meaning a "sharp or acute" fever) were common in the marshy areas. Geoffrey Chaucer (1342–1400) wrote in the *Nun's Priest's Tale*: "You are so very choleric of complexion/Beware the mounting sun and all dejection,/Nor get yourself with sudden humours hot;/For if you do, I dare well lay groat/ That you shall have the tertian fever's pain,/Or some ague that may well be your bane." And William Shakespeare (1564–1616) mentioned ague in eight of his plays. For example, in *The Tempest* (Act II, Scene II) Stephano mistaking Caliban's trembling and apparent delirium for an attack of malaria tries to cure him with alcohol "… he hath got, as I take it, ague … open your mouth: this will shake your shaking … if all the wine in my bottle will recover him, I will help his ague."

Quinine and Cinchona

In the sixteenth century, a serendipitous discovery led to a treatment for the ague. In the Viceroy's Palace in Lima, Peru, the beautiful Countess of Chinchon lay gravely ill with the ague. Her husband, the Count, fearful she would die, called the court physician to provide a remedy, but none was at hand. In desperation, the physician obtained a native Indian prescription: an extract from the bark of a tree growing in the Andes Mountains. The concoction was given to the Countess in a glass of wine, and the symptoms abated. The physician was rewarded, the Count relieved, and the Countess returned to Spain where she lived happily thereafter. The remedy that had been provided, and called by the Indians of Peru "quina-quina," literally "bark of barks," came to be known in Europe as the Countess' powder or the Countess' bark. This story of the Countess' recovery from her affliction—surely it must have been malaria that she had—circulated for 300 years in Europe; regrettably the story appears to be a fable. Who then first introduced fever bark into Europe? The most plausible explanation (and this is only a guess) is that the medicinal effect of the bark was discovered by the Spanish missionaries who came to Peru four decades after Pizarro's conquest of the Incas, and either by following the practices of local Indian herbalists or by trial and error its fever-curing properties were found.

Malignant fevers were so common in Europe during the seventeenth and eighteenth centuries that there was an increased demand for the powdered bark. The Jesuits through their missions had easy access to the bark; arranged for the collection of the bark in Peru, Ecuador, and Bolivia; powdered it, and then sold it in Europe for the benefit of the Society of Jesus. Because the Jesuit fathers were the promoters and exporters of the remedy, it was called Jesuits' bark or Jesuits' powder.

In the 1600s, Jesuits' bark was used almost everywhere in Europe; however, in one country, where the ague was a national calamity, it was shunned. The England of 1650 was "Puritan

England," and there was general prejudice against the Roman Catholic Church. Oliver Cromwell (1599–1658), the nation's Protector, was a zealous guardian of the Protestant faith who hated both the papacy and Spain. As a result, no one had the temerity to bring to England a medicine sponsored by the Vatican and known by the abhorrent name, Jesuits' powder. Although 2 years after the death of Cromwell (from malaria!) the first prescription of the powder in England was written the medicine did not become popular until 1682 when Robert Talbor's "secret for curing malaria" was disclosed.

Talbor, almost unknown today for his work with fever bark, was born in Ely in 1642 in the ague-ridden English fens and was determined to find a cure. He was not trained as a scientist, nor was he a member of the Royal Society of London, and he did not read or write in Latin as his medical contemporaries did. In 1661, Talbor was working as an apothecary's assistant and had access to Jesuits' bark, which he was quick to recognize its value if administered safely and effectively. He left his apprenticeship and conducted "field studies" on ague patients in the Essex marshes using different formulations and procedures. In 1672, he was appointed one of the King's Physicians in the Ordinary and styled himself as a specialist in fevers, and when in 1678 he was called to Windsor Castle and successfully cured the ague-suffering King Charles II, who knighted him for his service, Talbor's fame spread and in sympathy for his friends in France, King Charles allowed Talbor to visit the French court where he cured the Dauphin of his fever. The secret of Talbor's cure became a subject of intense interest and in 1679 the Dauphin's father, King Louis XIV, paid a large sum for the secret, provided it was not revealed until after Talbor's death. In 1682, following Talbor's death King Louis XIV published the remedy—the English "bitter solution." There was no real secret; rather, it was the simplicity of the method for administration and the dosage. He gave larger doses more frequently and rarely bled or purged his patients. The Jesuits' powder, infused with white wine to disguise its bitter taste, was sprinkled with the juices of various herbs and flowers and given immediately after the fit. One piece of information not revealed was which of the several different kinds of barks he used. Now the challenge for the rest of the world was to discover the kinds of bark that would have the greatest effectiveness against malaria.

The rising demand for the new remedy and a desire to better understand the trees that produced the useful bark led to a series of botanical expeditions to the New World to find trees in the wild that would ensure a predictable supply of high-quality barks. The earliest, in 1753, was sponsored by the French Academy of Sciences. The specimens were sent to Carolus Linneaus, the Swedish naturalist. Wishing to immortalize the name of the Countess of Chinchon, Linnaeus gave the tree the name *Cinchona*. However, in his *Genera Plantarum* (1742) and his *Materia Medica* (1749), he misspelled it, leaving out the first "h" of the Chinchon family name; despite the error, *Cinchona* remains enshrined as the name for the fever bark tree. Linnaeus prepared the first botanical description of two species, although only one, *C. officianalis*, had any fever-reducing properties.

Cinchona trees, of which there are 23 different kinds, grow in a narrow swath in cool climates on the slopes and valleys of the Andes; the trees do not grow lower than 2500 ft or higher than 9000 ft above sea level, and the forests are thick with hornets, mosquitoes, and vicious biting ants. The hardship of collecting the bark was considerable—the climate was variable; there was often a thick mist, sunshine alternated with showers and storms, and temperatures were near freezing. As a consequence, bark collection was relegated to the Indians, called cascadores or cascarilleros, who had found a clump of the desirable cinchona trees in the dense forest and proceeded to cut away the surrounding vegetation, removing the vines and parasitic plants that encircled the trunk; the bark of trunk was then beaten and longitudinal and circular cuts were made; and the tree was felled and the bark stripped. Slabs were dried over an open fire; the thickest parts were dried flat and the thin pieces from the branches were rolled into tubes. Both were packed into bales or put into sacks and transported down the torturous mountain trail to market. Until 1776, nearly all the bark collected was from *C. officinalis.*

The active component of the cinchona bark was unknown until Pierre Pelletier and Joseph Caventou isolated it in 1820. Pelletier was the son of a French pharmacist and had begun his work at the Ecole de Pharmacie in Paris and later became a retail apothecary in Paris. Caventou was another young Parisian pharmacist with a penchant for plant chemistry, who assisted Pelletier. The two began working on the bark in 1818 when Pelletier was 32 and Caventou 25. Both already had experience with other plant extracts including the isolation of strychnine and emetine. An alcoholic extract of the bark did not produce a precipitate when diluted with water; however, when alkali (caustic potash or potassium hydroxide) was added, it produced a pale yellow gummy substance that could be dissolved in ether, water, or alcohol, and they named the bitter-tasting alkaloid quinine after the Indian word "quina-quina." Pelletier and Caventou took out no patents for their discovery or the manufacture of quinine (as would be the common practice today); however, in 1827, the French Institute of Science rewarded them with a small sum of 10,000 francs. Pelletier began to manufacture quinine commercially in 1826, and in that same year a Swiss apothecary, Riedel, began its manufacture in Berlin for which he received $8 per ounce. In 1823, Farr and Kunzl (Philadelphia) manufactured quinine, and in 1824, Rosengarten & Sons (Philadelphia) began to manufacture the "essence" of the cinchona bark, actually quinine sulfate, a more soluble salt that had the advantage of being more easily swallowed than powdered bark and was less likely to induce vomiting. Both firms eventually merged with the German company Merck, and in 1837, it began to prepare quinine sulfate. This company was purchased by Engelmann & Boehringer, and in 1892, one of the Boehringer family's partners, Friedrich Engelhorn, became the sole owner, though he kept the name Boehringer. Boehringer became the biggest manufacturer of quinine and quinine products, and under Engelhorn's stewardship, a new company seal and trademark was designed, featuring a branch from the cinchona tree.

By the 1840s, export of the bark to Europe from the Andean republics amounted to millions of tons. The Spanish government stepped in and funded expeditions that found several new species of cinchona. In 1844, Bolivia passed laws that prohibited the collection and export of seeds and plants without a license because 15% of the country's tax revenue came from bark exports. The idea was to protect their monopoly, to discourage reckless stripping of the forests, and to prevent smuggling. In 1852, the Dutch sent a botanist to Bolivia and Peru, who narrowly succeeded in obtaining cinchona seeds that were used to start a plantation in Java in 1854. By 1850, the British had decided that a controlled supply of cinchona was necessary. The British Army in India estimated that it needed an annual supply of 1.5 million pounds in order to prevent 2 million adults from dying of malaria in India, and rehabilitating the 25 million survivors of malaria would require 10 times as much. And then there was Africa with a malaria fever rate up to 60% in some regions. In 1860, a British expedition managed to procure 100,000 seeds of *C. succirubra* that were used for plantings in British India and Ceylon. Because it had a quinine content of only 2% when compared to 17% from the red bark cinchona *C. calisaya* from Bolivia, it was a financial failure, and by 1900, the entire British venture was found to be unprofitable and a million trees were destroyed. Clearly, the seeds obtained by the Dutch and British were not of the best yielding varieties of cinchona growing in the Andes.

The most successful collector of cinchona seeds was a short, barrel-chested Englishman, a real cockney, Charles Ledger (1818–1905). At the age of 18, Ledger was working as a trader in alpaca wools in Peru. Ledger collected a herd of alpacas and walked them more than 1500 miles over the mountain passes from Bolivia to the coast of Chile where they were sent to Australia. However, the enterprise failed; the flock was decimated by disease, and finally broken up and sold leaving Ledger bankrupt. He returned to Peru making friends with the Indians and local traders; by living in the cinchona area, he was able to differentiate between the less and more active barks. Knowing of the world market for Jesuits' bark, he sent his servant Manuel Incra Mamani, a Bolivian cascarillero, to search for seeds from a stand of 50 huge *C. calisaya* trees that they had seen in flower in Bolivia 9 years earlier (1851) during the exploration of a mountainous passage for alpacas. Mamani was gone for 5 years because four April frosts destroyed the flowers and prevented the ripening of the fruit to produce seeds. But finally in 1865, he succeeded. He had walked more than 1000 miles from Bolivia to bring the seeds to Ledger. Sent out again by Ledger to collect more seeds, he was seized by the Bolivian officials who had banned the collection and export of the seeds. For refusing to tell for whom he was collecting, Mamani was imprisoned, beaten, starved, and robbed of his possessions. At last, he was set free only to die shortly thereafter from his ill treatment.

In 1865, Ledger sent 14 pounds of the high-quality seeds to his brother George in London, who attempted to sell them to the British government. The British government was not interested, and so the remaining half was sold to the Dutch for about $20. Within 18 months, the Dutch had 12,000 plants ready to set out and 5 years later their analyses of the bark showed the quinine content to be between 8% and 13%. To honor Ledger, this high-yielding species was named *C. ledgeriana*. Experimenting with hybrids and grafting onto suitable rootstocks, the Dutch developed the world's best cinchona trees—and used a process for obtaining the bark that did not destroy the tree. Ledger eventually retired to Australia where he received a miserable pension from the Dutch and in 1905 died in poverty; he was buried in a pauper's grave in Sydney. In 1994, a tombstone was erected in his memory bearing the inscription "Here lies Charles Ledger. He gave quinine to the world." There is, however, no monument to Manuel Incra Mamani.

The Dutch formed a cooperative, the Kina Bureau in Amsterdam, to control quinine production. Consequently, by 1918, 90% of the bark from Java, which represented 80% of the world's production, was sent to Amsterdam and distributed by the Kina Bureau. At the outbreak of World War II, Java had some 37,000 acres of cinchona producing more than 20 million pounds of bark a year. The Dutch quinine combine had created what amounted to the most effective crop monopoly of any kind in all history.

The Synthesis of Quinine

On December 7, 1941, the Japanese attacked Pearl Harbor, and as a consequence, the United States declared war. With the Japanese occupation of Java, the world's supply of quinine became limited (at the time 90% of the world's supply came from Java). With the shortage of quinine, there was a push by the U.S. military to synthesize quinine. News on the breakthrough in the synthesis of quinine by William Doering and Robert Woodward of Harvard University was heralded in the May 4, 1944, edition of the *New York Times* with the title: "Synthetic Quinine Produced Ending Century Search." The article went on to state: "the duplication of the highly complicated architecture of quinine molecule was achieved and ... considered it one of the greatest scientific achievement's of the century." And in the June 5th edition of *Life* magazine, an article appeared that was titled "Quinine: Two Young Chemists End a Century's Search by Making Drug Synthetically from Coal Tar." The *Science Newsletter* stated that starting with 5 pounds of chemicals, they had obtained the equivalent of 40 milligrams.

Although Woodward had promoted the synthesis of quinine beginning as early as 1942, his immediate aim was not for the use by the military, but for commercial purposes because he was supported by contracts from Edwin Land's Polaroid Corporation with the objective of finding synthetic alternatives to quinine as a precursor to light polarizing molecules. (Land was the inventor of instant photography using his innovative Land Polaroid camera.) Woodward and Doering's synthesis was not amenable to commercial production, however. Indeed, their strategy for synthesis would have cost 200 times more than the naturally derived product if, indeed, it was at all feasible. Moreover, it would have taken years of research to optimize the process and to reduce the price, and by that time there were alternative synthetic drugs.

Today, approximately 300–500 tons of quinine and quinidine are produced each year by extraction of the bark from cinchona trees. Approximately 40% of quinine is used in pharmaceuticals, whereas the remainder is used by the food industry as the bitter principle in soft drinks such as bitter lemon and tonic water. (Because tonic water contains only 15 mg of quinine per liter the drink has little antimalarial benefit.)

Quinine is still used occasionally in the treatment of severe falciparum malaria. Because of its slow action and rapid elimination, it is administered not by mouth but by slow intravenous infusion at a loading dose of 20 mg/kg body weight over 4 h, followed by maintenance doses of 10 mg/kg infused over 2 h every 8–12 h. If intravenous administration cannot be used, then it is given by intramuscular injection.

Despite its use for centuries, we still do not know precisely how quinine works, but we do know that it works only on the blood-feeding stages of malaria parasites. Malaria parasites have a unique apparatus (cytostome) for the ingestion of the red blood cell's hemoglobin, placing this in food vacuoles where by the action of protein-digesting enzymes the globin portion is broken down into smaller fragments (peptides and amino acids), releasing the potentially toxic-free heme that is polymerized into the inert, crystalline, golden brown-black, nontoxic, malaria pigment. It has been suggested that quinine acts within the food vacuole to block the crystallization of malaria pigment, thereby allowing the accumulation of large amounts of the toxic heme to accumulate and this kills the parasite.

The Disease Malaria

Malaria, literally meaning "bad air," was unknown in the English language until Horace Walpole in 1740 during a visit to Rome wrote: "There is a horrid thing called mal' aria that comes to Rome every summer and kills one." The disease was generally believed to result from human exposure to the poisonous vapors that had seeped in from the soil. In 1878, the French Army physician Alphonse Laveran was transferred to Bone, Algeria. The North African coast was malarious, and although Laveran spent most of his time looking at fixed material from those who had died from malaria, he also examined fresh specimens. On November 6, 1880, while examining a drop of fresh blood, liquid and unstained, from a feverish artilleryman, he saw several transparent mobile filaments—flagella—emerging from a clear spherical body. He recognized that these bodies were alive and that he was looking at an animal, not a bacterium or a fungus. Subsequently, he examined blood samples from 192 malaria patients: in 148 of these he found the telltale crescents. Where there were no crescents, there were no symptoms of malaria. Laveran also found spherical bodies in or on the blood cells of those suffering with malaria. He named the one-celled beast *Oscillaria malariae* and communicated his findings to the Société Médicale des Hôspitaux on December 24, 1880. Laveran was anxious to confirm his observations on malaria in other parts of the world, and so he traveled to the Santo Spirito Hospital in Rome where he met two Italian malariologists (one of whom was Ettore Marchiafava and the other Angelo Celli, Professor of Hygiene) and showed them his slides. The Italians were unconvinced. Later, Marchiafava and Celli claimed to have seen the same bodies as described by Laveran but without any pigment granules. They also denied the visit by Laveran 2 years earlier. Although the search for the parasite itself continued for most of this period, there was a serendipitous finding of great significance. In 1884, C. Gerhardt deliberately induced malaria for therapeutic purposes in two patients with tertiary syphilis by injection of blood from another patient suffering with intermittent fever, and then cured them all using quinine. A year later, Marchiafava and Celli, working on the wards of Rome's Santo Spirito Hospital, gave multiple injections of malaria-infected blood intravenously and subcutaneously to five healthy subjects. Parasites were recovered from three of the five who came down with malaria; all recovered after receiving quinine. Clearly, it was the blood of a malaria patient that was infectious, and the disease was not acquired from "bad air."

In 1886 during a visit to Europe, Major George Sternberg visited Celli at the Santo Spirito Hospital. Celli drew a drop of blood from the finger of a malaria patient and was able to show Sternberg the ameba-like movement of the parasite and the emergence of flagella. Sternberg returned to the United States and working with blood taken from a malaria patient in the Bay View Hospital in Baltimore was able to find Laveran's parasite in Welch's laboratory at Johns Hopkins University. A year later, Welch separated the two kinds of malaria with 48 h fever peaks; one would be named *Plasmodium* (from the Latin "plasma" meaning "mold-like") *vivax* and the other he named *P. falciparum* because it had sickle-shaped crescents (and "falcip" in Latin means "sickle or scythe"). In 1885, Camillo Golgi (1843–1926) of the University of Pavia examined 22 malaria patients with a quartan (72 h) fever cycle. Golgi discovered that the malaria parasite reproduced asexually by fission and correlated the clinical course of fever with destruction of the red blood cells to release the parasite. In 1886 when he noted that in both the tertian (48 h) and the quartan (72 h) fevers, there were no crescents; he effectively had distinguished *P. vivax* from *P. malariae* based on fever symptoms. Today, it is recognized that there are four kinds of human malaria: *P. falciparum, P. malariae, P. vivax,* and *P. ovale.* (Some investigators suggest that despite its preference for monkeys, *P. knowlesi* should be considered the fifth malaria parasite.)

All of the pathologic effects of malaria are due to the rapidly multiplying stages in the blood. The primary attack of malaria begins with headache, fever, anorexia, malaise, and aching muscles. This is followed by paroxysms of chills, fever, and profuse sweating. The fever spike may reach up to 41°C and corresponds to the rupture of the red blood cells as infectious stages (merozoites) are released. There may be nausea, vomiting, and diarrhea. Such symptoms are not unusual for an infectious disease, and it is for this reason that malaria is frequently called "The Great Imitator." Complications of malaria include kidney insufficiency, kidney failure, fluid-filled lungs, neurological disturbances, and severe anemia. In the pregnant female, falciparum malaria may result in stillbirth, lower than normal birth weight, or abortion. Nonimmune persons and children may develop cerebral malaria, a consequence of the mechanical blockage of small blood vessels and capillaries in the brain because of the sequestration of infected red blood cells.

All human malarias are transmitted through the bite of an infected female anopheline mosquito. A mosquito becomes infected when it ingests male and female gametocytes (that arise from merozoites in the blood that do not divide asexually). In the mosquito stomach, the male gametocyte divides into flagellated microgametes (first seen by Laveran) that escape from the enclosing red blood cell; these swim to the macrogamete; one fertilizes it; and the resultant motile zygote, the ookinete, moves across the stomach wall. This encysted zygote, now on the outer surface of the mosquito stomach, is an oocyst. Through asexual multiplication, threadlike sporozoites are produced in the oocyst, which bursts to release sporozoites into the body cavity of the mosquito. The sporozoites find their way to the mosquito's salivary glands where they mature, and when this female mosquito feeds again, sporozoites are introduced into the human body and the transmission cycle has been completed.

During the late 1930s and early 1940s, and before there was penicillin, patients with neurosyphilis were treated with malaria in order to cure them of the progressive dementia and paralysis (see Chapter 12). During malariatherapy, the effects of quinine treatment were found to be markedly different in blood-induced infections from those that were mosquito-induced, that is, by sporozoites. The blood-inoculated patients were cured with quinine, whereas the sporozoite-induced infections relapsed after the same quinine therapy. In order for there to be a relapse, malaria parasites must have been lurking somewhere in the body, but the question was where? In 1948, P. C. C. Garnham and H. E. Shortt at the London School of Tropical Medicine and Hygiene allowed 500 malaria-infected mosquitoes to bite a single rhesus monkey, then for good measure the infected mosquitoes were macerated in monkey serum, and this brew was also injected into the monkey. Seven days later, the monkey was sacrificed and its organs were taken for microscopic examination. Malaria parasites were found in the liver. From that time onward, liver (or exo-erythrocytic, EE) stages have been described for the human malarias, as well as many of the rodent and primate malarias.

In *P. falciparum*, the disappearance of infected red blood cells from the peripheral blood (as evidenced by simple microscopic examination of a stained blood film) may be followed by a reappearance of parasites in the blood. This type of relapse, called recrudescence, results from an increase in the number of preexisting blood parasites. *P. vivax* and *P. ovale* also relapse, although the reappearance of parasites in the blood is not from a preexisting population of blood stages and occurs after cure of the primary attack with a drug such as quinine. The source of these blood stages remained controversial for many years, but in 1980, the origin of such relapses was identified. In relapsing malarias, induced by sporozoites, Krotoski and coworkers found dormant parasites, called hypnozoites, within liver cells. The hypnozoites, by an unknown mechanism, are able to initiate full EE development and then go on to establish a blood infection.

The Road from Dyes to Antimalarial Drugs

After 1700, a complex series of social and economic changes took place in England and Germany; much of it was based on the development of the technology for using steam-powered engines. Coal was the cheapest and most abundant fuel used for powering these engines. At the same time, the coke oven was developed for heating coal to a high temperature in the absence of air to produce a viscous substance called coal tar. Coal tar was a complex carbon-containing substance consisting of a wide variety of organic molecules. Distillation of coal tar produced light oils and heavy oils known as creosote and the residue pitch. August Hoffmann (1818–1892), who in 1843 had just obtained his doctor's degree under Justus von Liebig at Giessen, Germany, was given the task of analyzing the contents of a bottle of light oil and found it contained benzene and aniline. In 1845, Hoffmann left Germany to become the director of the newly formed Royal College of Chemistry in London, an institution backed by industrialists and agriculturists. His appointment to the Royal College was favored by Britain's Prince Albert, with the hope that organic chemistry could be applied to improvements in agriculture. In 1854 Hoffmann assigned to William Perkin (1838–1907), then 16 years of age, the synthesis of quinine in the laboratory. Hoffmann had written: "Everybody must admit that the discovery of a simple process for preparing artificially the febrifuge principle of the cinchona bark would confer a real blessing on humanity. Now we have good grounds for the expectation that constructive chemistry will not long remain without accomplishing this task. Already … numerous substances have been artificially formed, which are in the closest relation to quinine …." Hoffmann believed the formula for quinine differed from that of an aniline derivative, allyltoluidine, by the addition of two molecules of hydrogen and oxygen and hence, he reasoned it should be possible to make quinine from it by just adding water through the action of potassium dichromate in sulfuric acid. At the time, chemists knew the number and kind of atoms in a molecule, that is, how many hydrogens, oxygens, and carbons there were, but not how they fitted together. Thus, the students at the Royal College conducted much of their experiments in organic chemistry without a map or compass. Indeed, if Perkin and Hoffmann had known the actual structure of quinine they certainly would have abandoned this route, but at that time the chemical structure of quinine and allyltoluidine was unknown. Perkin oxidized the allyltoluidine (prepared from coal tar with toluene) with dichromate and got a brown-red solid that was definitely not quinine. He repeated the experiment using "aniline," which actually contained a mixture of aniline and toluidine, and obtained a black precipitate. He boiled the black sludge with ethyl alcohol and a striking purple color solution formed, he called it aniline purple or mauveine. This was the first recorded preparation of a synthetic dye and Perkin quickly recognized its commercial value, because he was successful in using it to dye silk, and it did not fade or run when subjected to washing or when exposed to sunlight. From this accidental discovery, the synthetic dyestuffs industry began.

Due to the efforts of Hoffmann and his students, aniline dyes in all shades of the rainbow could be had. "The manufacture of the dye would prove to be crucial to developments in organic chemistry in that it shifted the attention of investigators from a consideration of color as an incidental property to one of primary importance. Thereafter, many chemists began to search consciously for organic compounds … with color." They could produce by artificial synthesis not only natural substances but those that did not occur in nature. Despite Hoffmann's admonitions that Perkin continue with his academic studies Perkin left school at 18, patented his discovery, built a factory and retired in 1873 (age 35) a rich man. Ironically, the poorly endowed Royal College of Chemistry lost its support because it could not show its usefulness to agriculture and British investors did not consider the support of scientific education and long-term industrial research as a good financial risk, especially when compared to investments in mining, trade, shipping, and the manufacture of textiles and railroad equipment. So in 1865 when the Prussian government solicited Hoffmann to return to Germany as the chair of chemistry in Berlin, a lucrative research post with large laboratories that he could design himself, he could not resist. With the departure of Hoffmann, the center of the dyestuffs industry moved from Britain to Germany, and from Berlin he would speak bitterly about the loss of financial support by the industry, the lack of encouragement for his teaching, and the British government's failure to appreciate the importance of chemistry as both a pure science and as a means to further industrial advancement. By World War I, about 90% of all dyestuffs were being manufactured in Germany.

In 1881, Paul Ehrlich (1854–1915) used methylene blue, a dye derived from aniline to stain bacteria. And 4 years later, he found that it also had a strong and specific affinity for living nerve fibers and other cells. Seizing on the fact that certain dyes stained only certain tissues and not others, it suggested to Ehrlich that there was chemical specificity of binding. This notion became the major theme in his scientific life and led to a search for medicines (derived from dyes) that would specifically strike and kill parasites. Ehrlich called these substances "magic bullets" and wrote that: "curative substances—a priori—must directly destroy the microbes provoking the disease; not by an 'action from distance,' but only when the chemical compound is fixed by the parasites. The parasites can only be killed if the chemical has a specific affinity for them and binds to them. This is a very difficult task because it is necessary to find chemical compounds, which have a strong destructive effect upon the parasites, but which do not at all, or only to a minimum extent, attack or damage the organs of the body. There must be a planned chemical synthesis: proceeding from a chemical substance with a recognizable activity, making derivatives from it, and then trying each one of these to discover the degree of its activity and effectiveness. This we call chemotherapy."

The first recorded cure of malaria by a substitute for quinine fell to Ehrlich. In 1889, after finding that methylene blue stained malaria parasites in the blood, Ehrlich administered capsules containing 100 mg dye five times a day to two patients who had been admitted to the Moabite Hospital in Berlin suffering from mild malaria (*P. vivax*). Both recovered, and although later methylene blue was found to be ineffective against a more severe kind of malaria (*P. falciparum*), this was the first instance of a synthetic drug being used against a specific disease. Ehrlich did not continue his work on the chemotherapy of malaria using methylene blue for two reasons. First, he did not have available laboratory animals that could be infected with malaria for testing potential medicines, a prerequisite for the development of any chemotherapeutic agent for use in humans or animals, and second, with his move in 1891 to the Institute for Infectious Diseases in Berlin he had to devote his attention to the preparation of vaccines (see Chapters 6 and 7).

Methylene Blue to Yellow Atabrine

In 1909, Wilhelm Roehl (an associate of Ehrlich) met Carl Duisberg (1861–1935), a chemist and a member of the board of management at Bayer's Elberfeld laboratories (and at that time a part of the IG Farben cartel) at a lecture in Frankfurt given by Ehrlich on the fundamentals of chemotherapy. Duisberg sought out Roehl with an eye toward bringing his expertise in chemotherapy to Bayer, and in 1911, Roehl responded to Duisberg's overtures. Roehl knew of Ehrlich's work on methylene blue and wrote: "it might have been surmised that by chemotherapeutic effort there might be discovered new compounds that would prove effective against malaria. Hence it is remarkable that excluding the cinchona alkaloids, and methylene blue ... no new therapeutic agents have been found heretofore." Roehl believed that a major stumbling block for malaria chemotherapy was the lack of a suitable animal model for studying the effects of putative drugs.

In 1911, P. Kopanis at the Hamburg Tropical Institute described the use of canaries infected with *P. relictum*, a malaria parasite isolated from infected sparrows and showed this could be used for testing the effects of quinine. Thus, in 1924 Roehl had, what Ehrlich lacked, a small laboratory animal able to be infected by malaria parasites. He noted that although Kopanis had been able to show that a drug active against human malaria was also active against this bird malaria, the reverse had never been done or as Roehl put it: "an active substance against malaria in birds which later proved effective against human malaria." In addition, Roehl identified another problem associated with the bitter tasting quinine: the subject might not accept the drug either in the food or in the water. Other researchers had overcome this drug-delivery problem by injecting the drug either into the muscle or under the skin; however, because in some instances this resulted in an immediate toxic or inflammatory response the dose had to be reduced significantly. When Roehl found that his canaries "would rather starve than swallow the bitter drug," he used a technique he had previously used at Speyer House. Using an esophageal tube—a fine catheter attached to a syringe—it was possible to quantitatively deliver a known amount of drug into the stomach of a 25 g canary. Initially, Roehl tested the effects of quinine on canaries that were infected with *P. relictum* by intramuscular injections. Normally parasites appeared in the blood in 4 or 5 days; however, if the drug dosage of quinine was high enough, it could delay the appearance of infected red blood cells for 10 or more days provided the canaries were treated daily for 5 days after inoculation of infected blood. To determine whether the drug was active, Roehl used untreated canaries as his controls and compared the delay in the appearance of parasites in the blood between the controls and the drug-treated birds.

Satisfied that he had a system that was "tractable, scalable, and quantifiable," Roehl moved beyond quinine to search for new antimalarial drugs. He had derivatives of methylene blue synthesized and found that substituting a diethylaminoethyl side chain on one of the methyl groups produced an effective cure of the malaria-infected canaries. He was concerned, however, that (as expected) a strongly colored dye might prove to be unacceptable to consumers. To avoid this problem, he switched to colorless quinolines but retained the basic side chain of the active methylene blue in the belief that this was essential for antimalaria activity (Figure 1.1). The strategy was to vary the point of attachment of the side chain to the quinoline to prepare a range of substitutions—hundreds, possibly thousands—and then to investigate the effectiveness of each in malaria-infected canaries. In 1925, one particular compound, a 6-methoxyaminoquinoline, was effective in curing canaries. We can imagine him shouting to all who could hear: "Only after my quantitative method had been elaborated was it possible to test the substances chemotherapeutically in animal experiments!"

In 1925, Franz E. Sioli, director of the regional insane asylum in Dusseldorf tested the 6-methoxyaminoquinoline, and the drug was found to be effective in curing the mental patients of their blood-induced vivax malaria. After further evaluation in clinics in Hamburg, Germany, and in naturally acquired infections in Spain and Italy, it was marketed worldwide as Plasmoquin. It was the first synthetic antimalarial and was sold throughout the world as a

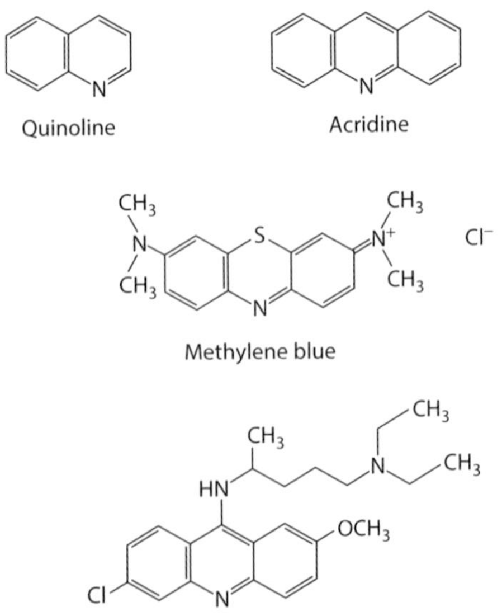

Quinoline   Acridine

Methylene blue

Atabrine

FIGURE 1.1   A comparison of the structures of quinoline, acridine, methylene blue, and the derivative of methylene blue, atabrine.

substitute for quinine. However, with further clinical trials Plasmoquin was shown to have many unfavorable side effects and therefore was not widely used.

In 1926, Robert Schnitzer at Hoechst (then a part of the IG Farben cartel) began to synthesize Plasmoquin analogs with an extra benzene ring added to the quinoline ring, thereby producing an acridine (Figure 1.1). In 1927, the project was taken up by Fritz Mietsch and Hans Mauss at Bayer in an attempt to synthesize novel acridines as possible antimalarials, because dyes (gentian violet, mercurochrome) were already in use as topical antibacterial agents. However, it was not the antibacterial properties of the yellow-colored acridine dyes that suggested their possible usefulness to treat malaria but the fact that they had a similarity to the antimalarial Plasmoquin. Using synthetic organic chemistry, Mietsch and Mauss tried to put the right substituents in the right positions so that, as Mietsch put it "they could bring out the slumbering chemotherapeutic characteristics."

More than 12,000 acridine compounds were synthesized and one, where the nitro group was replaced by a chlorine atom, was successful in antimalaria tests carried out by Walter Kikuth, who had replaced the recently deceased Roehl. By 1930, based on Kikuth's screening tests there was clear evidence of antimalarial activity by this chloroacridine; however, the first published reports did not appear until 1932. The drug was given the name Plasmoquin E, then changed to Erion, and later called Atabrine (Figure 1.1). In the United Kingdom it was given the name mepacrine.

In the early 1930s, Atabrine was introduced in the United States (where it was also called quinacrine) as a Winthrop Chemical Corporation product, but it was actually made in Germany by IG Farben; all Winthrop did was put it into ampoules or to compress it into tablets for distribution under its own label. Atabrine was marketed throughout the world as a substitute for quinine; however, it was not without problems including mental disturbances. Winthrop neither conceded the fault nor mentioned it. They were satisfied with releasing "friendly" favorable statements and took comfort in the tremendous expansion made at the outbreak of World War II for defense purposes. In addition, the selling price was reduced to a figure that was one-tenth of that charged before the war. Paul de Kruif, author of *Microbe Hunters*, wrote in *Readers' Digest* that Winthrop's Atabrine was a major victory for the United States never once mentioning its untoward side effects. It was, however, the best available drug for curing malaria especially after the Japanese controlled the world's supply of quinine by its invasion of Indonesia. The United States responded to the Japanese attack on Pearl Harbor by ordering Winthrop to supply large amounts of Atabrine for use by the military in the South Pacific where malaria was as great a threat as Japanese bullets. Prior to that Winthrop had merely produced 5 million tablets annually from six chemical intermediates imported from Germany. (Winthrop had been set up in the United Sates after World War I to distribute Bayer pharmaceuticals, following its purchase by the Sterling Drug Company of New York a subsidiary of IG Farben. In 1926, by a subsequent agreement IG Farben took over Bayer and became half owners of Winthrop. After the attack on Pearl Harbor, a U.S. government antitrust suit severed the ties between IG Farben and Winthrop so the latter became a wholly owned American company.) By 1944, Winthrop through a sublicensing and royalty-free agreement with 11 American manufacturers was producing 3500 million tablets of Atabrine. Because the antimalaria activity of Atabrine was similar to quinine, that is, it killed the parasites growing in red blood cells but did not affect the parasites in the liver it could be used prophylactically to suppress the symptoms of malaria.

During World War II, it was recognized that a victory by the United States would be tied to the development of a number of research and development programs. The most famous of these was the Manhattan Project, which produced atomic bombs; the Radiation Laboratory that developed radar; and a crash program for the development of antimalarials. It began in 1942, under the auspices of the National Research Council (NRC), it was a massive program coordinated by a loose network of panels, boards, and conferences. The actual work was done at universities, hospitals, and pharmaceutical industry laboratories as well as in U.S. Army and U.S. Navy facilities. At Johns Hopkins University, canaries, chicks, and 60,000 ducks were infected with malaria parasites and used to screen and test 14,000 potential antimalarial compounds for activity. At the Illinois State Penitentiary in Joliet and in Atlanta (Georgia), New York, and New Jersey federal and state prisoners were used as human guinea pigs: after they were subjected to the bites of malaria-carrying mosquitoes the men were given potential antimalarials to determine effectiveness. The focal point for the clinical evaluation of drugs emerging from the various animal screening programs was the Research Service of the Third Medical Division of Goldwater Memorial Hospital in New York City, under the direction of James A. Shannon (1904–1994).

In 1941, Shannon was a young assistant Professor of Physiology at the NYU School of Medicine and his area of expertise was the mechanism whereby the kidneys form urine (renal physiology) not malaria. Shannon had developed ways to monitor the effect of hormones and various drugs on kidney function and had worked during summers at the Mount Desert Island Biological Laboratory with E. K. Marshall, Professor of Pharmacology at Johns Hopkins, and Homer Smith a world famous kidney physiologist. Shannon had joined Smith's NYU lab in 1931 after completing his residency at Bellevue Hospital, where he remained for 9 years, earning a PhD to go with his MD. Marshall was now a prime mover on one of the NRC boards, the Board for the Coordination of Malaria Studies, responsible for screening compounds with the potential for curing malaria and to provide a better understanding of Atabrine's worth and limitations. Marshall had helped develop two sulfa drugs and he had pioneered a new and more quantitative approach to dosage setting. Paul Ehrlich would have been proud of Marshall! Until then, the method of setting the dose for a medicine was crude at best and entirely empirical. A patient was given a drug of a given dosage and afterward the symptoms of the disease were observed. Atabrine was given at the approved dose of one-tenth of a gram three times a day, an amount based on that used with quinine. Shannon was critical of this unscientific method and wrote: "Such an approach to the general problem of Atabrine therapy is a striking contrast to the more quantitative one which has facilitated the development of sound antibacterial therapy with sulfanilamides." What was needed, Marshall believed, was someone to determine the effective dose of Atabrine because the troops were not inclined to take it. It made them sick, turned their skin and eyes yellow suggesting that it might cause liver damage leading to jaundice, and in many instances it not only failed to suppress malarial attacks it was painfully slow acting when the fever–chill paroxysm did come. Atabrine was rumored among the soldiers

FIGURE 1.2   A sign from the South Pacific theater of operations.

to impair sexual vigor, a belief the Japanese military exploited through air dropped leaflets (Figure 1.2). The Americans responded with billboards showing a jolly sultan leering at a dancing girl as he popped a pill, remarking "Atabrine keeps me going!"

The key to the new approach by Shannon's team at Goldwater was to find a way to measure the concentration of the drug in the blood plasma. Why? Because blood (and urine) is the most accessible body fluid and it is close to what a particular tissue "sees." Once the level in the plasma could be determined then it would be possible to measure the levels with varying doses. Today, this seems obvious but in the early 1940s few appreciated its significance for successful treatment. By the spring of 1943 as General Douglas MacArthur was about to send his troops into the South Pacific; Shannon, Bernard (Steve) Brodie, and a laboratory technician Sidney Udenfriend working in the basement of Goldwater had solved the problem using atabrine's intrinsic fluorescence, a property some soldiers had already discovered when they urinated in the moonlight!

Using measures of fluorescence, the amount of atabrine was determined easily and routinely in the blood, and the method also worked with urine, feces, or any body tissue. When an experiment was carried out with a dog the body tissues, particularly muscle and liver, were found to soak up atabrine. This explained why the drug was slow acting. The solution: increase the dosage, but doubling or tripling the dose might kill not only the malaria parasites but could exacerbate the intolerable side effects. Shannon's way around this problem was to initially provide a big loading dose and then to follow this with relatively small daily doses to maintain adequate blood levels. This is what was done, and by 1944 it could be said: "Malaria as a tactical problem had practically disappeared." Shannon left Goldwater in 1946 to become the director of the Squibb Institute for Medical Research, and there he helped develop streptomycin and expand its production. Soon however, Shannon grew disenchanted with the pharmaceutical industry. He felt too tainted by the high life, by too much money and was distressed in seeing that research was unduly distorted by commercial considerations. He left Squibb for the

National Institutes of Health (NIH), first as head of the National Heart Institute and eventually he became the director of the NIH. When he first arrived at NIH, he said that he had hoped to return to research one day, but he never did. Instead, Shannon once a bench scientist became a great bureaucrat and research administrator and as director (1955–1968) oversaw the spectacular growth of NIH.

Chloroquine

On April 12, 1946, an article in the *New York Times* declared, "Cure for malaria Revealed after 4-year, $7,000,000 research. The curtain of secrecy behind which the multimillion dollar government anti-malaria program had been operating, in the most concentrated attack in history against this scourge, was completely lifted today for the first time, with the revelation of the most potent chemicals so far found." The next day, a *New York Times* editorial stated: "When the scientific story of the war is written, we have here an epic that rivals that of the atomic bomb, the proximity fuse and radar." The drug receiving so much attention was chloroquine. In the United States, it was the result of the screening of some 14,000 compounds (with each given a survey number, SN) under the aegis of the NRC's Board for the Coordination of Malarial Studies.

The most important of these new compounds was SN-7618 (Figure 1.3) prepared by German chemists, discarded by the Germans, tested by the French, and rediscovered by the Americans. In 1934, chemists at Bayer's Elberfeld laboratories synthesized colorless

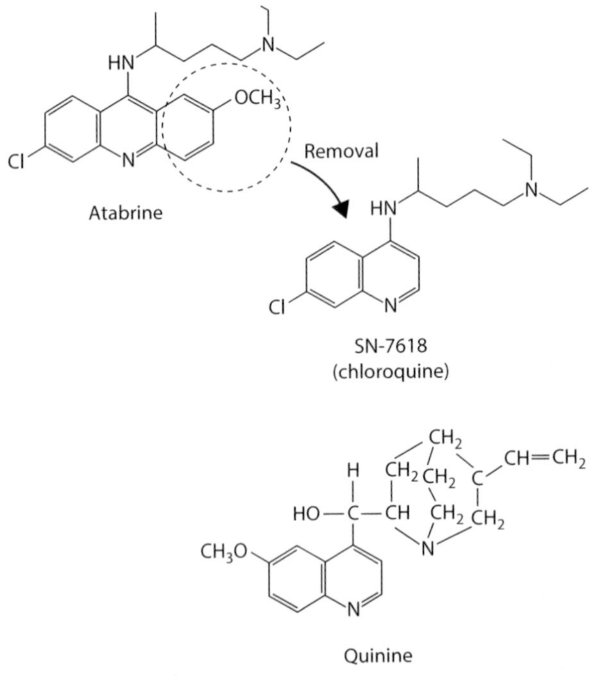

FIGURE 1.3   A comparison of the structure of chloroquine with quinine and atabrine.

4-aminoquinoline and made two salts, one of which, 2, 4-dihydrobenzoic acid, was named Resochin (because by German terminology it was the RESOrcinate of a 4-aminoCHolin). When screened using Roehl's bird malaria test, it was found to be as effective as atabrine. It was also effective against blood-induced *P. vivax* infections. However, based on tests in animals it was considered too toxic for practical use in humans and so Bayer abandoned it. Undeterred, in 1936, Bayer chemists produced a methylated derivative of Resochin, called Sontochin. This was less toxic and just as effective as atabrine. Bayer patented both Resochin and Sontochin, and through the IG Farben cartel (of which Bayer was a member) the same patents were taken out in the United States in 1941 by Winthrop Chemical Company (also a member of the cartel). To gain patent protection in France, where there was no protection for production of medicines, Bayer sent samples of Sontochin to a French pharmaceutical company, which in turn sent it to Tunisia for testing in humans. In early 1942 Sontochin was found to be effective against *P. vivax* and without adverse reactions. In late 1942, when the Allies invaded North Africa the French investigators offered the remaining supplies of Sontochin to the U.S. Army. This material "which had been captured from the enemy in North Africa" was then tested as SN-183 by the Division of Chemistry and Chemical Technology of the Board for the Coordination of Malaria Studies and found to be identical to the compound (Sontochin) that Winthrop had made under manufacturing directions received from IG Farben in 1939. Although the patent was not issued until 1941, it had already been tested and found to be effective in malaria-infected canaries by Maier and Coggeshall at the Rockefeller Foundation Laboratory in New York. Despite this, Winthrop, for some reason, put SN-183 on the shelf where it was forgotten. In 1943 when the data on SN-183 was reviewed by the Division of Chemistry and Chemical Technology, the Chairman (E. K. Marshall!) made the mistake of considering its structure to be that of an 8-amino-quinoline, and wrote: There is no need to study additional 8-aminoquinolines. Thus, SN-183 was dropped from further testing. Later, when it was discovered that Winthrop's SN-183 was identical to Sontochin, SN-183 was dropped as a designation, and a new SN number, SN-6911, was substituted. SN-6911 was found to be four to eight times as active as quinine in the bird malaria *P. lophurae*; however, in humans it was no more effective than atabrine. And, by this time, American chemists had produced a more effective 4-aminoquinoline, a 7-chloro derivative (=SN-7618).

In dogs, SN-7618 was nine times as toxic as Sontochin; however, in monkeys it was one-fourth as toxic, and when in 1944 it was tested for toxicity on conscientious objectors at the Massachusetts General Hospital in Boston no adverse toxic symptoms were observed. Further trials in malaria therapy were conducted at the Boston Psychiatric Hospital. Another plus: SN-7618 was without the undesirable side effects of atabrine. After being successfully tested against several bird malarias, further clinical trials were conducted. SN-7618 persisted longer in the blood so that it could be used prophylactically, and a single weekly dose of 0.3 g was able to maintain a blood plasma level for complete suppression of malaria. Harry Most at the NYU Medical Center showed that 1.5 g given for 3 days was effective against acute attacks of *P. vivax* as well as *P. falciparum*. Clearly, SN-7618 was the drug of choice for the management of malaria and it was given the name chloroquine. In late 1944, the board was informed that chloroquine was not patentable as a new antimalarial because its structure was identical to Resochin, and already covered by two patents owned by Winthrop Chemical Corporation. Large-scale production of chloroquine made atabrine obsolete.

The story of the discovery of chloroquine (marketed under the trade names Aralen, Nivaquine) makes clear that the reliance on the Roehl *P. relictum* test and the convenience and ease of using bird malarias can be misleading unless the proper host–parasite combination is used. And as the Johns Hopkins biochemist William Mansfield Clark (Chairman of Committee on Chemotherapy=Division of Chemistry and Chemical Technology of the Board for Coordination of Malaria Studies) observed that it would seem wise to pay less attention to the toxicity of a drug in the bird and more attention to the toxicity of the drug in mammals. Indeed, although bird malarias were essential to the World War II effort aimed at drug development for use in humans, it became apparent there were limitations to their utility and there was a need for mammalian models.

A respected malaria researcher, Robert Desowitz, in his book *The Malaria Capers*, wrote: "Chloroquine spread like a therapeutic ripple throughout the tropical world. In stately homes, the bottle of chloroquine would be a fixture, along with the condiments on the family table. In military cantonments the troops would assemble for a 'chloroquine parade'. In hospitals and rural health centers the malarious of all skin colors were treated with chloroquine … ." In 1955, the World Health Organization (WHO) confident in the power of chloroquine to kill malaria parasites and DDT to kill the malaria-carrying *Anopheles* mosquitoes launched the Global Malaria Eradication Campaign. As a result, malaria was eliminated from Europe, Australia, and other developed countries and large areas of tropical Asia and Latin America were freed from the risk of infection but the campaign had excluded sub-Saharan Africa. The WHO estimated that the campaign saved 500 million human lives that would otherwise have been lost due to malaria, but the vast majority of the 500 million lives saved were not Africans.

However, in the 1960s there were ominous reports of treatment failures with chloroquine. By 1969, many of the countries where malaria was endemic became eradication weary and there was a resurgence of disease. Many of these countries saw no end to the demands on their funds with ever diminishing returns on their investments in DDT and chloroquine. *Anopheles* mosquitoes had developed enzymes to detoxify DDT and also changed behaviorally—not settling on the walls that had been sprayed with DDT instead flying into homes, biting, and then leaving. In addition, there were strains of falciparum malaria in South America and Southeast Asia that were no longer susceptible to the killing effects of chloroquine and these were now spreading across the globe. The WHO gave way slowly and grudgingly and began to endorse a policy of control. By 1972, the Global Eradication Program was formally declared dead, but even before this it was recognized that another "magic bullet" to replace chloroquine was desperately needed to attack the disease in the half billion sick people living in some 90 countries or territories.

Mefloquine

Malaria parasites have been toughened by decades of exposure to antimalarial drugs—conditions that promote their survival and to resist the assault by "magic bullets." In this cauldron, where evolution does not stand still, the parasites are constantly developing countermeasures to defend themselves against the drugs that have been designed to kill them. In a sense, the drug war against malaria is effectively an arms race between a killer (*Plasmodium*) and survival of its victim (*Homo sapiens*).

In the 1960s, it was clear that the honeymoon for chloroquine was over. Although the Vietnam War began in 1959, and the first American combat troops did not actually arrive in South Vietnam until early 1965, by 1962 it was already apparent to the U.S. military that the malaria situation in the country was serious. Unlike malaria during the Korean War (1950–1953) the prevalent parasite in Vietnam was not *P. vivax* but the deadly *P. falciparum*, and worse still it was drug-resistant. By 1963, the incidence of malaria in U.S. troops in Vietnam had risen to 3.74 cases per 1000 compared with a worldwide figure for military personnel of 0.13 per 1000. Brigadier General Robert Blount, Commanding General, U.S. Army Medical Research and Development succinctly stated the case: "Once more, malaria ranks as a top military medical priority for the Department of Defense." It was estimated that 1% of the U.S. soldiers were acquiring malaria for every day they spent in combat in South Vietnam despite receiving chloroquine. By 1969 there were 12,000 cases of malaria in the troops in Vietnam with a loss of 250,000 man-days and direct medical costs of $11 million; yet between 1959 and 1969, the U.S. government was spending less than 2% of its $486 million for malaria on basic research.

Therefore, the development of a drug to replace chloroquine would not be driven by a strategic necessity as with past shortages of quinine but rather by the military's need to combat drug-resistant *P. falciparum*. There was another constraint: a reluctance on the part of the pharmaceutical industry to develop new drugs for malaria because the investments of time, money, and resources to market a new medicine were so high it was unlikely that a reasonable return on the investment was possible. Thus, protection and treatment of the U.S. armed forces led the U.S. Congress to expand funding for research into malaria. In 1963, Colonel William Tigertt the director of Walter Reed Army Institute of Research (WRAIR) set into motion the machinery for the U.S. Army Research Program on Malaria in what was to prove to be the largest screening program ever undertaken with a mission to identify and develop antimalarial agents effective against the emerging drug-resistant strains of *P. falciparum*.

From 1970 onward, hundreds of quinoline methanols were examined for antimalarial activity in *Aotus* monkeys infected with *P. falciparum*. Twelve specially selected 4-quinoline methanols were tested with chloroquine-resistant and chloroquine-sensitive *P. falciparum*, and from this screening five derivatives were as active or more active than chloroquine against strains susceptible to this drug and equally effective against strains that were resistant to chloroquine, quinine, and pyrimethamine. WR-142,490, examined in the *P. falciparum–Aotus* monkey model was the most active and it emerged as the flagship response to chloroquine-resistant

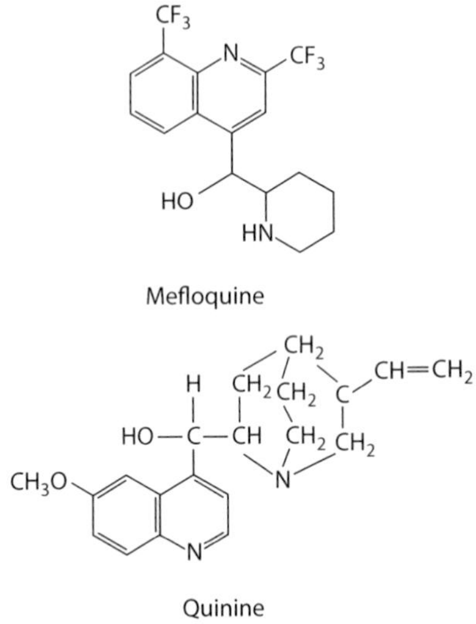

FIGURE 1.4   A comparison of the structure of mefloquine with quinine.

P. falciparum. The administration of a single dose was as effective as the same amount of drug divided into 3 or 7 doses over as many days. Intravenous administration of the drug in monkeys was also feasible. In human volunteers, when taken orally, the drug was safe and effective at 1500 mg for 1 day or in 500 mg weekly for 52 weeks. Peak concentration of the drug in the blood was 12–36 h and the mean half-life was 14 days. Nonimmune volunteers developed no *P. falciparum* infection after a single dose and in those infected when treated there was rapid clearance of fever and blood parasites in the chloroquine-sensitive and chloroquine-resistant strains. WR-142, 490 was named mefloquine.

Mefloquine is structurally related to quinine (Figure 1.4), is selective in killing asexual stage parasites and does not have activity against gametocytes or liver stages. Mefloquine and quinine are both lipophilic ("lipid loving") and bind tightly to serum and high-density lipoproteins and this may facilitate entry into the parasite. Mefloquine also binds to the membranes of the red blood cells and accumulates in the parasite food vacuoles, as does chloroquine. However, it is not clear if both drugs share the same mechanisms of action.

The Fall of Mefloquine

In an intensive and fruitful collaboration, the WHO sponsored more than 12 clinical trials of mefloquine in Latin America, Zambia, and Thailand. And in 1979, Hoffmann-LaRoche launched the drug by itself (as a monotherapy) under the trade name Lariam (or Mephaquine); it was licensed in the United States in 1988. Lariam was the drug of choice for travelers and visitors to areas where chloroquine-resistant malaria was present.

It was suggested by Wallace Peters (London School of Hygiene and Tropical Medicine) as early as 1977, using a variety of rodent malarias, that emergence of resistance to mefloquine would be reduced if administered in combination with another antimalarial. Indeed, Peters wrote: "It is strongly recommended that mefloquine should be deployed only for the prevention or treatment of malaria in humans caused by chloroquine-resistant *P. falciparum*. Although it is appreciated that data based on rodent malaria models may not necessarily predict the situation in human malaria, the authors suggest that, for large scale use, mefloquine should not be employed until a second antimalarial has been identified that will minimize the risk of parasites becoming resistant to this potentially valuable new compound." Indeed, in the early 1980s the first reports of resistance to mefloquine appeared. This prompted the WHO in 1984 to issue a publication expressing reservations concerning the widespread use of the drug and suggested that it be used in combination with mefloquine with another antimalarial; this the WHO said might preserve the effectiveness of mefloquine by delaying the emergence of resistance as well as to potentiate its activity. Hoffmann-La Roche had produced such a formulation by combining mefloquine with Fansidar (marketed as Fansimef). Indeed, simultaneous administration of mefloquine with sulfadoxine/pyrimethamine (Fansidar) delayed the emergence of mefloquine resistance in the *P. berghei* mouse model and when *P. falciparum* isolates were grown in the laboratory cultures. Clinical trials with Fansimef (mefloquine combined with sulfadoxine/pyrimethamine (250/500/25 mg, respectively) carried out between 1982 and 1986 in Africa showed similar effectiveness against falciparum malaria as mefloquine on its own. However, in Southeast Asia particularly on the Cambodian border with Thailand and Vietnam some clinicians questioned the wisdom of the use of mefloquine in fixed combination with Fansidar in countries where there was already resistance to sulfadoxine/pyrimethamine.

There were further complications with the prophylactic use of mefloquine: neuropsychiatric episodes, including insomnia, strange or vivid dreams, paranoia, dizziness or vertigo, depression, and attempts at suicide presumably due to a blockade of ATP-sensitive potassium channels and connexins in the substansa nigra in the brain. The overall risk varies with ethnicity (highest in Caucasians and Africans than Asians) as well as differences in health, cultural, and geographical background so it is not certain what are the actual reasons for the differential adverse reactions. Due to concerns over the safety of mefloquine prophylaxis in Western countries the packet insert has been revised (July 2002) stating: "Use of Lariam (mefloquine Roche) is contraindicated in patients with known hypersensitivity to mefloquine or related compounds such as quinine and quinidine. Lariam should not be prescribed for prophylaxis in patients with active depression, generalized anxiety disorder, psychosis or other major psychiatric disorders or with a history of convulsions." For these reasons as well as the development of mefloquine resistance in endemic areas in parts of Southeast Asia and the loss of efficacy in such areas, exemplified the danger of introducing a long-acting antimalarial where there is high transmission and where other control strategies are blocked. Therefore, "it is unlikely that mefloquine alone will ever regain its clinical efficacy in such areas."

Sweet Wormwood and Qinghaosu

The report in the December 1979 issue of the *Chinese Medical Journal* (*CMJ*) announcing "An effective antimalarial constituent … extracted from a Chinese medicinal herb … named Qinghaosu" surprised some members of the Steering Committee on the Chemotherapy of

Malaria (CHEMAL), a component of the WHO Special Programme for Research and Training in Tropical Diseases (TDR) meeting in Geneva. The impetus for establishing TDR began in May 1974 when the World Health Assembly called for a program to "intensify activities in the field of research on the eight major tropical parasitic diseases." By 1975, the TDR was committed to "strengthen research and training activities particularly in developing countries." Although initiated by WHO, by 1979 TDR had several sponsors—the UNDP (United Nations Development Program) and the World Bank. Scientists and public health experts were brought together to help set the Programme's overall research priorities and goals. The operations of TDR were overseen by specialized Scientific Working Groups (SWGs), each led by a Steering Committee, one of which was CHEMAL. CHEMAL had its first meeting in the summer of 1976. When the Chinese announcement was made in the *CMJ*, Walther Wernsdorfer and Peter Trigg, the Secretariat for CHEMAL, and some Steering Committee members were already aware of the progress that had been made in the People's Republic of China on qinghaosu (pronounced as "ching how sue"). Indeed, in the fall of 1978, Walther Wernsdorfer and William Chin had conducted a workshop on the in vitro cultivation of *P. falciparum* and its application to chemotherapeutic and immunological research at the National Serum & Vaccine Institute in Beijing, where qinghaosu was used as a model compound for studies on drug sensitivity. However, at the time of the report in the *CMJ*, CHEMAL was dealing with the development and registration of new drugs related to mefloquine (see p. 23) in collaboration with WRAIR and Hoffmann-LaRoche.

Excited by the news the CHEMAL members asked themselves: Was it possible that the best hope on the horizon for curing malaria was to come from a 2000-year-old drug or was it another instance of the Chinese claiming "as in the past, most conspicuously, when they said they could cure malaria by acupuncture only to be proven wrong later after there were clinical trials?" Some believed the acupuncture reports, dating back a decade or more, and suggested this was proof that the Chinese could not be trusted in this case either. By late 1979, skeptical not of the science but perhaps of the willingness of the Chinese to collaborate on the development of a novel antimalarial drug, the WHO arranged for some of its staff to visit China and the TDR agreed to conduct training courses on drug sensitivity using a modification of the in vitro culture system for *P. falciparum* that had been developed 3 years earlier by Rockefeller University scientists William Trager and James Jensen. China also agreed to a technical visit by the Chairman of CHEMAL (Wallace Peters) to review the Chinese antimalarial drug development program. In another visit, Peter Trigg conducted the training course at the Institute of Parasitic Diseases in Shanghai, and on behalf of CHEMAL, Trigg and Peters had discussions on the status of qinghaosu at the Center for Traditional Medicine at the Second Military College Institute in Beijing. From these discussions it became apparent that the Chinese were making and growing qinghaosu and wanted to supply and sell it. What was unclear was the degree and type of collaboration that the Chinese required from the outside world. They clearly wanted assistance in upgrading their facilities to GMP (good manufacturing practices) and the design of international acceptable studies but whether they wanted cooperation in production and marketing was another matter. The discussions reflected the political uncertainty in China at that time regarding international cooperation and capitalization in a communist state with Deng Xiaping as leader of the group that wanted a capitalist approach within communism and the hardliners (military and others) who were wary of cooperation particularly with the United States. There was a period in the early 1980s when things moved forward slowly until the student troubles and Tiananmen Square when things began to cool off. The Chinese Ministry of Foreign Affairs sought the help of WHO in establishing collaborative arrangements; however, it was apparent that there was disagreement between the military and the clinicians who had used the drug in human trials and felt there was no need for further toxicity and dosing studies and those scientists who understood the Western approach to drug development which included extensive preclinical studies. The impression gained from this visit was that the chemistry carried out by the Chinese was excellent; however, preclinical studies were largely nonexistent. In 1980 a WHO Scientific Working Group on the development of qinghaosu and its derivatives was held in China to review the data and to develop a plan of action. The Chinese authorities agreed that CHEMAL should ask WRAIR to assist in preclinical development, and in turn WRAIR agreed to develop protocols. In 1982 another visit was made by WHO (TDR) and together with a representative from the FDA the production facilities in Kunming and Guilin were reviewed as well as the clinical studies carried out in Quangchou (Canton). None of the facilities were found to meet GMP standards, so the FDA spent time training staff on the necessary principles and practices of GMP. Although the Chinese wanted the drug to be sold and used, they were uncomfortable at TDR taking over the development work. TDR fully recognized the rights of the Chinese to develop qinghaosu (in English artemisinin); however, validating safety and efficacy of a drug for the improved treatment of malaria worldwide would require a broader testing effort.

Preclinical development by CHEMAL required one kilo of artemisinin; however, even after several more visits to China this was not forthcoming. To obtain the raw material, Arnold R. Brossi (1923–2011), a member of CHEMAL and Chief of Natural Products Section in the NIH Laboratory of Structural Biology (who held a PhD in chemistry (1952) and had been at Hoffmann-LaRoche from 1953 onward eventually becoming director of its Chemical Research Department (1963–1975) arranged for 500 g of artemisinin to be produced under the auspices of CHEMAL by a grower and extractor in Argentina. Ostensibly, Brossi's idea was to produce a less toxic derivative; however, he had an ulterior motive: to provoke the Chinese into providing a kilo of artemisinin. The ploy worked. One day in 1983 Trigg was summoned to the office of the Chinese Assistant Director General at WHO and was presented with a package containing a kilo of artemisinin. When he asked, "Whom should I thank?" the reply was "No one. It should not be acknowledged and CHEMAL should proceed to do with it what they wished."

Artemisia to Artemisinins

The story of the discovery of artemisinin had its basis in the political climate in China. During the Vietnam War (1959–1975), the Chinese government supported the Vietnamese Communists against the United States, which was becoming increasingly involved in the fighting. Malaria was rampant in Vietnam causing casualties on both sides, and Ho Chi Minh asked Mao Zedong for new antimalarial drugs to assist the Vietnamese troops in their fight against the imperialists. On May 23, 1967, the China Institute of Sciences set up a top secret military program called "Project 523" with the express purpose of antimalarial drug discovery that could be used in the battlefield. Project 523 involved over 500 scientists in ~60 different laboratories and an additional 2000–3000 people were involved at one time or another. Efforts in the first 2 years included screening ~40,000 compounds for activity but these yielded disappointing results. The project then looked at Chinese herbal preparations. One of the herbal preparations, taken from an ancient Chinese recipe book (*Recipes for 52 Kinds of Diseases*) written in 168 BC, was used as a treatment for hemorrhoids, and there is also a reference to it in the *Handbook of Prescriptions for Emergency Treatment* written by Ge Hang (284–346 AD). Ge Hang gives instructions to those suffering from intermittent fevers to soak a handful of the leaves of the sweet wormwood plant, *Artemisia annua*, in about a liter of water, squeeze out the juice, and drink the remaining liquid. In 1596 Li Shizen, the great Chinese herbalist, wrote in his *Compendium of Materia Medica* that the chills and fever of malaria can be combated with preparations made from *A. annua*.

Professor Youyou Tu joined Project 523 in 1969 and she and her research group examined more than 2000 Chinese herbals that might have antimalarial activity, and more than 200 recipes from Chinese traditional herbs and 380 extracts from these herbs were tested in a rodent malaria model. In 1971 Tu and her group identified qinghaosu (translated as "artemisinin") meaning active principle of Qinghao (English name *Artemisia annua*) from dried *A. annua* using low heat and ethyl ether extraction after they discarded those parts of the herbal extract that made it sour. In May 1972, the date of the first official report, it states that when the preparation was fed to mice (50 mg/kg daily for 3 days) infected with *P. berghei* malaria, it killed asexual stage parasites and was 95%–100% effective as a cure. Monkeys infected with sporozoites of *P. cynomolgi*, and after the appearance of parasites in the blood, were given 200 mg/kg of qinghaosu daily for 3 days via gastric tube and the blood was cleared of infected red blood cells in 2–3 days but there were relapses. Qinghaosu had no effect on liver stage parasites in *P. cynomolgi* infection. Tu and her colleagues bravely tested the active compound on themselves before knowing its chemical structure and 7 months later, in 1972, there was another report on the successful treatment in Beijing of 30 patients with malaria and over 90% of them were said to have recovered either from infections with *P. falciparum* or *P. vivax*. It was only then that a systematic characterization of the chemistry and pharmacology of the active compound, artemisinin, began.

When in 1973, Youyou Tu (later a 2015 Nobel laureate in Physiology or Medicine) and her colleagues synthesized the compound dihydroartemisinin (DHA) to prove that the active material had a ketone group; they were unaware that DHA would be more effective than the natural substance and only later would it be shown that DHA is the substance produced in the body after ingestion of artemisinin acting to clear the malarial fever. Although DHA, the metabolite of artemisinin, is far more active than its parent compound, the activity of artemisinin itself is also notable. Moreover, artemisinin has a much longer half-life and maintains the metabolic conversion to DHA over a relatively long space of time. In 1975, Professor X. T. Liang found

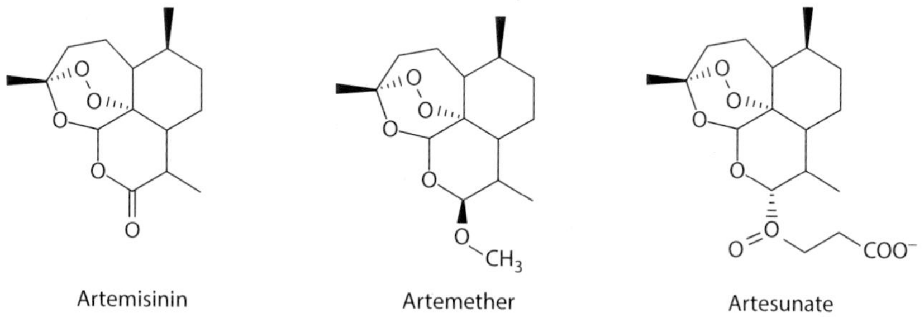

Artemisinin   Artemether   Artesunate

FIGURE 1.5   A comparison of the structures of artemisinin, artemether, and artesunate.

that qinghaosu with the empirical formula $C_{15}H_{22}O_5$ contained a stable peroxide in a complicated compound, a sesquiterpene lactone bearing an endoperoxide located at one side of the molecule in almost a semi-spiral form.

Between 1975 and 1978, clinical trials with artemisinin (Figure 1.5) on more than 6550 patients with malaria were successfully achieved. Artemisinin is volatile, and poorly soluble in water. In China, new variants of artemisinin were developed such as the methyl ester artemether (Figure 1.5), soluble in oils and suitable for oral administration, and the hemisuccinate ester artesunate (Figure 1.5), which is more soluble than artemether and can be used for intravenous injection. The Qinghaosu Antimalarial Coordinating Research group proceeded to treat and document the therapeutic response of *P. vivax* and *P. falciparum* infections in humans with artemisinins. Artemisinin was effective in chloroquine-resistant cases and produced a rapid recovery in 141 cases of cerebral malaria. There was no evidence of serious toxicity; however, there was a high relapse rate. Artemisinin works quickly and is eliminated quickly. After oral or intravenous administration artemisinin is converted to DHA in the body. The DHA is eliminated with a half-life of ~1 h. By 1981 the work on artemsinin was presented to the WHO/CHEMAL Scientific Working Group and in 1982 it entered the mainstream literature.

In 2001, the global demand for artemisinin skyrocketed when the WHO recommended artemisinin as the standard treatment for malaria. But the world's supply of artemisinin has been unreliable. *Artemisia annua*, mostly cultivated in China and Vietnam, requires up to 10 months from sowing to harvest, and the yield and quality vary depending on weather, region, growing practices, and market conditions. In 2006, Jay Keasling at UC Berkeley was able to successfully microbially engineer small quantities of artemisinin, and with a grant from the Bill & Melinda Gates Foundation, PATH pulled together Keasling's team and researchers from the biotechnology start-up Amyris, Inc. to turn the semi-synthetic laboratory production of artemisinin into a platform suitable for industrial development. The challenge was to find a way to scale up the process and expand production from just 2 liters of artemisinin at a time—about the size of a large soda pop bottle—enough to fill a vat the size of a three-story building. By 2007 the team had perfected a method: they took five genes from the *A. annua* plant and inserted them into yeast. The yeast ferments and produces an intermediate product called artemisinic acid. Using a chemical process this is then converted into artemisinin. To achieve large-scale production of the semisynthetic artemisinin, a partnership was established with the pharmaceutical company

Sanofi to optimize and scale-up the fermentation process for artemisinic acid. Using an innovative method that uses light instead of chemicals, the artemisinic acid is converted into artemisinin. Sanofi expected to make about 35 tons of artemisinin in 2013 and 60 tons in 2014, which would meet about one-third of the global need.

How does artemisinin do its dirty work to kill the malaria parasite? It has been shown that in the parasite heme (in the form of hemozoin) catalyzes the activation of artemisinin to irreversibly bind to as many as 124 parasite proteins especially key metabolic enzymes. Resistance of *P. falciparum* to artemisinin has been shown both in laboratory experiments and in the field and this resistance results from a mutation (Kelch13), which prolongs the length of time the parasite spends as a ring stage. Because artemisinin activation is low at the ring stage parasites with the Kelch13 mutation are able to overcome protein damage; thus they are selected as they have a higher capability to survive the drug treatment at the ring stage. On the other hand, the much higher level of drug activation at the latter developmental stages (trophozoite and schizont) triggers extensive protein modifications that act like an exploding bomb, inhibiting multiple key biological processes eventually resulting in parasite death. Thus, it is less likely that the parasite will develop resistance at the latter stages.

ACT Yes, Resistance No

The worsening problem of antimalarial drug resistance in many parts of the world coupled with the limited number of antimalarials available has provoked both a search for new drugs as well as the development of guidelines to protect the effectiveness of available drugs by limiting the emergence of resistance. Observations since the 1960s have shown where there was increased mortality from malaria it was directly related to the continued use of increasingly ineffective antimalarials such as chloroquine. Thus, there has been considerable concern that resistance in the field would emerge to the artemisinins as it has to almost every other class of antimalarials. It has been possible to select strains of rodent malarias that are five to ten times as insusceptible to artemisinin and by in vitro selection the same level of resistance has been accomplished with *P. falciparum*. After 2001, the use of artemisinins as a single drug for treatment decreased significantly to prevent the emergence of resistance. More importantly, resistance to artemisinin began to appear in isolated areas where it had been used on its own. In 2006, the WHO requested discontinuance of the manufacturing and marketing of all artemisinin monotherapies except for the treatment of severe malaria.

It was found that when artemisinin was combined with a partner drug to which malaria had not already become resistant, the new combinations were effective and well tolerated, although they were more expensive than the failed single drug treatment. Further, the rapid clearance of artemisinin drugs in treating early cases of uncomplicated malaria may prevent progression to severe disease and reduce mortality. Because of their very rapid clearance from the blood complete cure using artemisinins requires longer courses of therapy: given alone a 7-day regimen is required to maximize cure rates. Adherence to a 7-day course of treatment, however, is frequently poor so the combination partner drug in an artemisinin combination therapy (ACT) is usually a more slowly eliminated drug. This allows a complete treatment course to be given in 3 days. This being so, in 2001 a WHO Expert Consultative Group endorsed ACT as the policy standard for all malaria infections in areas where *P. falciparum* is the predominant infecting species. The WHO endorsed ACTs for the treatment of malaria in 2004, and recommended a switch to ACTs as a first-line malaria treatment in 2005.

In 1994, Novartis formed a collaborative agreement with the Chinese Academy of Military Sciences, Kunming Pharmaceutical Factory, and the China International Trust and Investment Corporation to further develop artemisinin combinations and eventually it registered Coartem and Riamet. Coartem constitutes about 70% of all current clinically used ACTs. Coartem (Novartis) consisting of artemether/lumefantrine, 20/120 mg, in tablet form was approved by the FDA in April 2009. It is highly effective when given for 3 days (6 doses) and with a small amount of fat to ensure adequate absorption and thus efficacy. In ACT, the artemisinin component acts quickly over a 3-day course of treatment and provides antimalarial activity for two asexual parasite cycles resulting in a staggering reduction of the billion blood parasites within an infected patient yet some parasites would remain for the partner drug to remove (and this is variably assisted by an immune response). Therefore, "the artemisinin component of the ACT reduces the probability that a mutant resistant to the partner drug will arise from the primary infection, and if effective, the partner should kill any artemisinin-resistant parasites that arose."

Aspirin box. (Courtesy of the Science Museum, London, Wellcome Images.)

Chapter 2

The Painkiller, Aspirin

In 1757, the Reverend Edward Stone of Chipping Norton in Oxfordshire, England, stripped the bark from a willow tree (*Salix alba*), and when it was tasted, he was surprised at its bitterness. It resembled, he thought, the taste of the Jesuits' bark, the rare and rather expensive remedy for the fever, the ague, a disease then quite common in the low-lying areas of England, particularly Oxfordshire and the fens of Kent. For the next 6 years, Reverend Stone examined the properties of the willow bark as a cure for the ague (malaria). On June 2, 1763, when Stone was convinced that the willow bark was a cheap and effective substitute for quinine, he read a paper titled "An Account of the Success of the Bark of Willow in the Cure of Agues" before the Royal Society in London. In the paper, it was stated, "There is the bark of an English tree, which I have found by experience to be a powerful astringent, and very efficacious in curing aguish and intermittent disorders. It seems likewise to be a safe medicine." Stone also suggested that the willow bark worked wonders in reducing fever because it was based on the traditional doctrine of signatures that held that "many natural maladies carry their cures along with them, or their remedies lie not far from their causes." Because willow trees grew in the same damp low-lying marshy areas that were malarial, it seemed plausible for Stone to postulate that this was the rationale for the willow bark being an effective remedy for ague (malaria).

Stone's remedy was to take dried white willow bark, grind it with a mortar and pestle, and administer it to ague sufferers. He "gave it in small quantities … about 20 grains of the powder at a dose and repeated it every 4 hours between fits … the fits were considerably abated but did not entirely cease. Not perceiving the least ill consequences, I grew bolder with it, and in a few days increased the dose to two scruples, and the ague was soon removed. It was then given to several others with the same success, but I found it better … when a dram of it was taken every 4 hours in the intervals of paroxysm." By the time Stone wrote to the Royal Society, he had "treated 50 persons and never failed the cure except in a few autumnal and quartan agues, with which the patients had been long and severely afflicted." Stone's letter, published in the *Philosophical Transactions*, received little attention except for a 1798 report by an English pharmacist named William White who wrote: "Since the introduction of this bark into practice in the Bath City Infirmary and Dispensary as a substitute for the Cinchona, not less than 20 pounds a year have been saved to the Charity."

The irony of Stone's discovery was that he did not find a cure for the ague that was as effective as quinine. What he had found was a therapeutic remedy for malaria's symptoms—fever, high temperature, aching limbs, and headaches. The real significance of Stone's work was that he had stumbled on an extraordinary substance that could relieve all these symptoms and cause no harm. However, it was not until a half century later when, driven by national rivalry, French and German pharmacists competed to identify the active fever-reducing principle found in willow bark.

Willow Bark to Aspirin

In 1828, Joseph Buchner, Professor of Pharmacy at Munich University, obtained a tiny amount of bitter-tasting crystals from willow bark and named the substance salicin (because *Salix* is the generic name of the willow). A year later, the French chemist Henri Leroux refined the extraction procedure and obtained an ounce of salicin crystals from 3 pounds of bark; and in 1838, Raffaele Piria of Pisa produced an acid from salicin, salicylic acid. Karl Lowig of Berlin was also able to obtain salicylic acid from the meadow sweet flower (*Spiraea ulmaria*), and he named it "spirasaure"; when he tested on himself and volunteers, he found it reduced fever and pain.

In 1860, Hermann Kolbe and his students at Marburg University went a step further and synthesized salicylic acid (Figure 2.1) and its sodium salt from phenol, carbon dioxide, and sodium. One of those students, Friedrich von Heyden, then established a factory in Dresden to produce salicylates on an industrial scale. This allowed production at one-tenth the price of the natural product. At the von Heyden plant, thousands of pounds of salicylic acid were produced. In 1874 Thomas McLagan in the Dundee Royal Infirmary in Scotland treated patients suffering with painful arthritic symptoms due to rheumatic fever with salicin. He wrote up the results of his rigorous trials, and these were published in the journal *Lancet* on March 4, 1876. He stated: "salicin lessened the fever and inflammation and pain." (He did not, however, cure rheumatic fever that is caused by a streptococcus.) As a result of these clinical studies, other physicians began to publish the results of their studies with salicin as well as salicylic acid; some workers claimed that salicyclic acid was better and that it could also help with headache, migraine, and neuralgia. Between 1877 and 1881, four of the main teaching hospitals in London were using salicylates as a regular therapy. In 1875, Germain Sée introduced salicylates as effective treatments for gout and arthritis.

By the 1870s, the German synthetic dye industry was preeminent worldwide. Realizing the potential of the new synthetic dyestuffs, Johann Weskott and Friedrich Bayer established a dye-producing factory in 1863. Bayer died in 1880, and when Weskott died a year later, the company was headed by Bayer's son-in-law Carl Rumff, who renamed the company the Dye Factory formally known as Friedrich Bayer and Company. Rumff recruited young chemists to conduct dye research, and one of the most talented of them was Carl Duisberg (1861–1935) who had received a doctorate in 1882. Initially, Duisberg's task was to synthesize blue and red dyes, and after succeeding, he was then asked to find new areas into which the company could expand. Duisberg learned that Ludwig Knorr at one of Bayer's rival dye companies had stumbled on a fever-lowering drug, supposedly a synthetic substitute for quinine, named antipyrine. Antipyrine (also known as Phenazone) was not a substitute for quinine; however, it did have a modest commercial success as an analgesic. Duisberg also learned that another chemical called acetanilide had been produced by acetylation of aniline, and this too had fever-reducing properties. Because there was no secret in making acetanilide, it was trademarked by Kalle & Co. as Antifebrine. This was a revolutionary move by a pharmaceutical company because it enabled physicians to use a generic term for a drug rather than a complex chemical name. Unfortunately, Antifebrine had a short-lived commercial success because its side effects were bone marrow depression and anemia.

Duisberg wondered whether Bayer might also produce a fever-reducing drug similar to Antifebrine but without the side effects. He gave the task to a doctoral student, Oskar Hinsberg, who produced acetophenatidine starting with aniline; it too was an antipyretic and had less harmful effects. Mindful of the power of trademarking, Duisberg named it Phenacetin. (Both

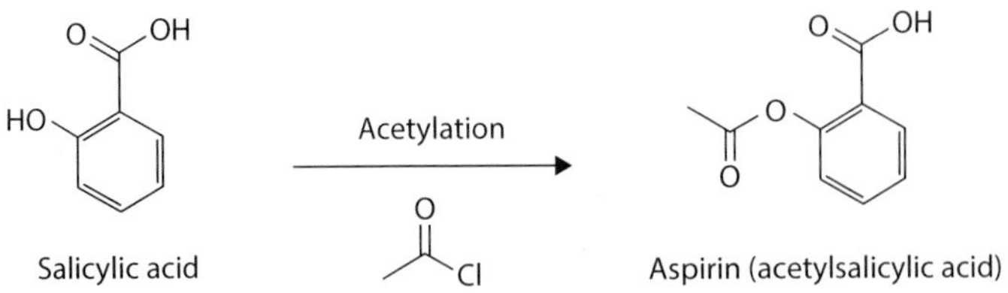

FIGURE 2.1 Formation of acetylsalicylic acid (aspirin) by treatment of salicylic acid with acetic anhydride.

acetanilide and phenacetin break down in the body to form *N*-acetyl-*p*-aminophenol and, by various anagrammatic combinations, yields the generic name acetaminophen and is today marketed by the trade name Tylenol.) Phenacetin made Bayer a great deal of money, and the success of phenacetin suggested to the entrepreneurial Duisberg that Bayer should set up a discrete pharmaceutical division with proper laboratories for chemists, and by 1890, the Bayer laboratory in Erlangen was operational and organized into two sections: one was a pharmaceutical group that came up with ideas for new drugs and a pharmacology group for testing these. The head of pharmacology was Heinrich Dreser, and his opposite number in pharmaceuticals was Arthur Eichengrunn. In 1896, the flamboyant, charismatic, and talented chemist Eichengrunn assigned a junior chemist, Felix Hofmann, to synthesize a drug that acted like salicylic acid but did not have the side effects: a foul taste, irritation of the stomach lining leading to vomiting and ulceration of the stomach, and did not cause tinnitus. Hoffman treated salicylic acid with acetic anhydride and produced acetylsalicylic acid (ASA) (Figure 2.1). ASA still had the sour taste of salicylic acid but was not corrosive to the stomach. Eichengrunn tested the ASA on himself and sent small quantities to doctors and dentists in Berlin and received glowing assessments of its pain-relieving properties. ASA, he concluded, was a general-purpose analgesic without unpleasant side effects. The ASA was then handed over to Dreser for clinical testing. Believing it enfeebled the heart as did salicylic acid, Dreser did not give it his seal of approval. However, after learning that Eichengrunn had gone behind his back, Dreser was incensed, and he tried again to kill the drug's release but Eichengrunn appealed to Duisberg. Duisberg ordered another full set of clinical trials and was pleased with the findings. On January 23, 1899, a memo was circulated among the senior management at Bayer posing the question: How should ASA be trademarked? Because salicylic acid had been obtained from the meadow sweet plant *Spiraea*, it was suggested that this should be part of its name; the letter "a" could be added in front to acknowledge acetylation and the letters "in" added on the end to make it easier to say. Bayer introduced it to the market under the brand name "Aspirin," because the German Patent Office had previously refused a patent for ASA.

Although defeated Dreser did his duty and wrote a prelaunch paper, in a fit of pique, he omitted to mention the contributions of Hoffmann and Eichengrunn. As a result of there being no patent on Aspirin, Eichengrunn and Hoffmann received no royalties on its sales, but Dreser had negotiated a special deal that ensured he was paid royalties on all medicines tested in his laboratory. Thus, Dreser became very rich and was able to retire early. Within 15 years of its release, aspirin was one of the most widely used drugs in the world providing Bayer with millions of dollars.

The Darker Side of Aspirin

In the 1920s, Duisberg as the head of Bayer had a vision: put similar dye makers under a single management and supervision structure with one sales organization to remove competition by many different products. In 1925 the vision was realized when six dye manufacturers merged to form a single cartel, IG Farben. It was the largest corporation in Germany, and as the dye market began to shrink, the cartel diversified its portfolio of products. In 1933, an association was forged between the National Socialist Party (Nazi) and IG Farben when Chancellor Adolf Hitler and Reichstag President Hermann Goering convinced the company that Germany was faced with a threat of economic decay and Bolshevism. Initially, IG Farben contributed money derived from its pharmaceuticals (principally aspirin) to support the political aspirations of the National Socialists, but later when the Nazis helped to rescue the company from a failing business venture to make synthetic gasoline the ties between the party and the company became stronger. By the start of World War II, IG Farben, at the zenith of its industrial might, was critical to Hitler's military goals in providing strategic materials—synthetic oil, rubber, and nitrates (for explosives). When the Nazis needed Zyklon B gas for use in the gas chambers to provide a final solution of the Jews, an IG Farben subsidiary Degesch provided it. By the mid-1940s, IG Farben was an integral component of the Third Reich, and it had fully acceded to Hitler's demands of total commitment. The cartel provided monies for the construction of Auschwitz; ran a chemical plant, IG Monowitz, where thousands of slave laborers worked and died; and it financed the medical experiments conducted by Nazi doctors, the most infamous being Josef Mengele. IG Farben went along with the Aryanization of its workforce and purged most of its scientists who were Jewish. One of those purged was Arthur Eichengrunn who, after leaving Bayer and establishing a new company Cellon-Werke to produce among other things cellophane, was forced in 1938 to sell the company. In 1943, Eichengrunn was deported to the concentration camp Theresienstadt. The camp was liberated by the Soviet Army in 1943, and only then was he able to tell the full story of his part in the discovery of aspirin.

In 1948, 23 IG Farben executives were tried for war crimes at Nuremberg, and 13 were acquitted by managing to convince the judges that they had been acting under duress. Among those convicted were the chairman of the IG Farben board and the senior production manager of IG Monowitz factory who were sentenced to 7-year imprisonment. After Nuremberg IG Farben was broken up, and three new companies arose from its ashes: Hoechst, BASF, and Bayer. Bayer reverted to producing pharmaceuticals as it had done when Carl Duisberg was in charge, and aspirin, the drug that started it all, continued to be one of its most profitable and successful products.

How Aspirin Works

H. O. J. (Harry) Collier (1912–1983) was the first to propose a mechanism for the action of aspirin. Collier, a British subject born in Rio de Janeiro, was a bright student and won a scholarship to Cambridge University where he earned a PhD. He initially worked on testing penicillin for the armed forces and then had positions at pharmaceutical companies, notably the London branch of Parke-Davis where he set up a pharmacology department. In 1958 he began experiments with guinea pigs to see if he could discover the mode of action of aspirin. It had recently been discovered that damaged cells released kinins into the blood, and when these attached to adjacent nerve endings, it caused pain and inflammation. Because quantifying pain levels was impossible, Collier used another approach. He administered a particular kinin, bradykinin, which caused the air passages of a guinea pig lung to constrict. Using this measurable effect, he began giving aspirin before and after administering bradykinin. If the aspirin was given before (but not after), the air passages of the guinea pig remained open. Collier realized that to nail down the role of aspirin he needed help. He recruited a bright, young pharmacology student to Parke-Davis, Priscilla Piper. For the next 5 years, they attempted to define the mechanism of action of aspirin using live guinea pigs, rabbits, and rats; however, they failed.

Collier had heard of a technique being used by John Vane (1927–2004) at the Royal College of Surgeons in London that might help to define the biochemical mechanisms, and so he asked Vane to take Priscilla Piper on as a graduate student to teach her the technique expecting that she would return to his lab. Vane accepted her; however, instead of returning to Collier's lab Piper switched camps, and together with Vane they soon discovered how aspirin worked.

Vane's assay, called the cascade superfusion bioassay, involved immersing two strips of tissue into a flowing solution, one upstream from the other. The first tissue, usually guinea pig lung, would be injected with a substance to induce anaphylactic shock, and the damaged tissue would then secrete a substance that was carried downstream to the second piece of tissue, usually rabbit aorta, to see its reaction (twitching). Using this assay, Vane and Piper found that adding aspirin prevented the second piece of tissue from reacting to the chemical substance released from the first tissue. They called it the rabbit aorta contracting substance (RCS). What was RCS chemically? In the 1950s and 1960s, Sune K. Bergstrom in Sweden had shown that, when a tissue is injured, it released arachidonic acid, and this in turn resulted in the production of hormone-like fatty acids called prostaglandins that had previously been discovered by his student Bengt I. Samuelsson. (The name prostaglandin comes from the word "prostate" because they were first isolated from human semen and thought to be associated with the prostate gland.) One weekend in April 1971, Vane had a flash of insight. He asked himself: "What if RCS was a prostaglandin? And what if aspirin prevented the release of prostaglandin?" The following Monday Vane did the experiment and found that aspirin inhibited the production of prostaglandin, and this he surmised would in an intact animal halt the fever, inflammation, and pain. On June 23, 1971, Vane and Piper published their results in the prestigious journal *Nature* with the title: "Inhibition of prostaglandin synthesis as a mechanism of action for aspirin-like drugs." Collier had been scooped.

In the following years, a great deal more would come to be known about aspirin's action. Aspirin blocks the production of the cyclooxygenase enzymes 1 and 2 (COX-1 and COX-2) that catalyze the conversion of arachidonic acid to prostaglandin. Two prostaglandins, in particular E and F, are responsible for causing inflammation, including redness and fever. And when platelets (which contain only COX-1) are taken from the volunteers given aspirin, they fail to make prostaglandins in response to the clotting factor, thrombin. In activated platelets, thromboxane, a platelet aggregator and a constrictor of blood vessels, is produced by the enzyme thromboxane synthetase. Endothelial cells contain primarily COX-2 that may be induced by the shear stress as blood flows through the blood vessels, and it too is blocked by aspirin. (Aspirin, by acting to reduce both thromboxane production and platelet clumping, lowers the chance of a myocardial infarction, i.e., a heart attack.) By 1974, Vane and Sergio Ferreira had convincing evidence that almost all aspirin-like drugs (called NSAIDs, nonsteroidal anti-inflammatory drugs) such as naproxen (trade name Aleve) and ibuprofen (trade name Advil) inhibit both cyclooxygenases. NSAIDs also act to reduce the side effects of stomach irritation because they block the synthesis of prostaglandins that the stomach lining needs to regulate the overproduction of acid and to synthesize the mucus barrier that prevents self-digestion.

In 1975, Bergstrom and Samuelsson confirmed that RCS was actually thromboxane. In 1982, Vane, Bergstrom, and Samuelsson were awarded the Nobel Prize for Physiology or Medicine, "for their discoveries concerning prostaglandins and related biologically active substances." The other heroes of the aspirin story were Collier who left Parke-Davis in 1969 and went to work for Miles Laboratories, which ironically was taken over some years later by Bayer; and Piper who though becoming a celebrated scientist in her own right died in 1995 of cancer.

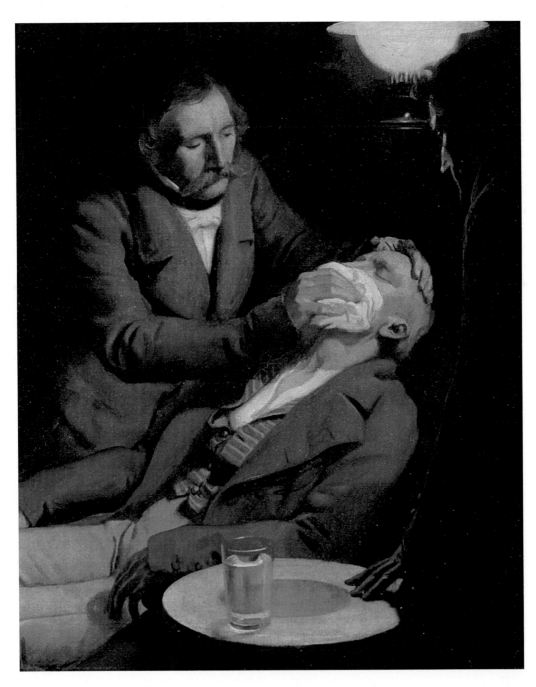

William Morton using ether in 1846. This is a painting by Ernest Board. (Courtesy of the Wellcome Library, London.)

Ether, Chloroform, Cocaine, Morphine, Heroin, and Anesthesia

During the Middle Ages and until the nineteenth century, cleanliness was neither a priority of medicine nor a priority of people in general. Communal wells provided water to the streets, but there were few facilities for removing sewage. Waste polluted the wells. The streets were dumping grounds for garbage and human wastes, and domestic animals, including pigs, roamed the streets. Uneaten food was thrown on the floor to be devoured by dogs and cats or to rot, drawing swarms of flies from the stable. The cesspool in the rear of the house would have spoiled a person's appetite if the sight of the dining room had not.

Surgery was an unsafe practice, because control of infection and anesthesia were unknown. In general, surgery was restricted to surface cutting, and opening of the body cavity was avoided, because it usually meant sure disaster. There were two kinds of surgeons: "true" surgeons, who carried out major operations such as removal of tumors, repair of fistulas, operations on the face, and amputations, and the barber surgeons, who would assist surgeons by carrying out bloodletting, extraction of teeth, resetting of bone fractures, and removal of skin ulcers. Gunshot wounds and injured limbs, frequently gangrenous, were amputated and then treated with cautery or boiling oil. The settings for early practitioners of the "cutting art" were the kitchen table, the battlefield, or below deck on a ship and not the hospital or the operating theater. Hospitals of the nineteenth century, poorly ventilated and overcrowded, were not institutions for the care of the sick as they are today, but they existed primarily to provide a refuge for the poor, orphans, and the insane (see Chapter 12).

Anesthesia

Before the 1800s, the methods for minimizing pain during surgery were few—opium, nightshade, strapping, compression of nerve trunks, distracting noises, and hypnotism. In addition, among the surgeons, the notion existed that anesthesia was a threat to their profession, stating "that it was unworthy to attempt to try by artificial sleep to transform the body into an insensitive cadaver before beginning to use the knife." Some even claimed that the pain endured during surgery was critical to recovery. The best surgeons were speed artists. Indeed, one of the most famous was London surgeon Robert Liston (1794–1847). Liston was famous for his nerves of steel, his satirical tongue, and an uncertain temper. He was renowned for his speed and dexterity during the performance of an operation. It was said that when he performed an amputation, the gleam of his knife was followed so instantaneously by the sound of the saw so as to make the two actions appear almost simultaneous. Liston removed a 45 lb scrotal tumor in 4 minutes, but in his enthusiasm, he cut off the patient's testicles as well. He amputated a leg in 2.5 minutes, but the patient died shortly thereafter from gangrene (as was often the case in those days). In addition, he amputated the fingers of his young assistant, who also died later from gangrene, and slashed the coattails of a distinguished surgical spectator, who was so terrified that the knife had pierced his vital organs that he dropped dead in fright. This was the only operation in history with 300% mortality. However, in the early 1840s, attitudes toward pain began to change, but this came from dentists, not from surgeons. Why were dentists so inclined to use agents that were able to produce insensibility to pain? Some have suggested that unlike surgeons, who would have contact with their patients only once (and sometimes with fatal consequences), dentists would have to minimize pain in order to keep their patients coming back.

Laughing Gas

The first dental anesthetic to be used was nitrous oxide (N_2O) or "laughing gas." In 1798, Sir Humphrey Davy (1778–1829) performed research on N_2O, a gas that had been identified by Joseph Priestly in 1772. He found that severe pain from dental inflammation would be relieved by breathing the gas and correctly observed "that it is capable of destroying pain, (and) it may probably be used with advantage during surgical operations." Yet, very few European surgeons used it. They believed that the pain endured during an operation was beneficial to the patient. However, in the United States, N_2O was used to amuse and demonstrations accompanied lectures for the purpose of entertainment. The American dentist Horace Wells (1815–1848) attended one of these lecture demonstrations. He saw the possibilities and wrote: "Reasoning by analogy I was led to believe that surgical operations might be performed without pain." He borrowed some of the gas and had a fellow dentist extract one of Wells's own wisdom teeth painlessly during inhalation. There followed dozens of extractions by using "painless dentistry," and with the help of another dentist, William T. G. Morton (1819–1869), Wells arranged to demonstrate the anesthetic properties of laughing gas at Harvard University. In January 1845, Wells addressed the senior class on his theory of painless dental extraction and then proceeded to demonstrate. One of the Harvard students volunteered to have his tooth removed, but because the bag of laughing gas was removed too early, he screamed from excruciating pain. Wells was shattered by the experience. He became the laughing stock of the medical community and left the practice of dentistry to engage in a variety of odd pursuits, including the selling of canaries and fake paintings and etchings.

In 1847, Wells began self-experimenting with the newer anesthetic, chloroform. The experiments that he conducted eventually led to him becoming addicted to chloroform, and he began to suffer from bouts of depression and hallucinations. During one of his delusional fits, he threw sulfuric acid on two prostitutes on Broadway and was promptly arrested and imprisoned in the Tombs in New York City. While in prison, he managed to obtain a bottle of chloroform and a razor. On January 24, 1848, under the influence of the chloroform, he cut the femoral artery in his left thigh and bled to death. He was only 33 years old. A few days after his death, the Paris Medical Society recognized Wells's contribution to general anesthesia and to perform painless surgery, and today, in a park in Paris, there stands a statue of Wells that commemorates his discovery. In 1875, Wells's hometown (Bushnell Park) erected a statue in his honor. At its base are a walking stick, a book, a scroll, and a bag (for N_2O). However, the practice of N_2O anesthesia was abandoned until J. H. Smith, another dentist, 10 years later, anesthetized several hundred patients and removed 2000 teeth painlessly. By 1886, N_2O was in common use in dentistry in the United States, where it is still used today. N_2O works as an anesthetic because it inhibits *N*-methyl-D-aspartate receptors (not gamma aminobutyric acid, GABA) of the nervous system, which play a role in learning, memory, and pain perception.

Ether

Ether (named by August Sigmond Froebenius in 1730) is a distilled mixture of sulfuric acid and alcohol, and in 1605, Paracelsus described its properties: "… if it is taken even by chickens … they fall asleep from it … but awaken later without harm. It quiets all suffering without any harm and relieves all pain and quenches all fevers. …" In 1818, Michael Faraday (1791–1867), who was working in Sir Humphrey Davy's laboratory, reported the effects of inhaling ether. He found that after inhalation of ether vapors, the lethargic state was followed by a great depression of spirits and a lowered pulse—conditions that might threaten one's life. In England, John Snow (who later achieved fame for his theory of cholera transmission) was among the first to witness the use of ether for tooth extraction and was captivated by the gas's power to produce "quietude" in the patient. However, unlike others, Snow did not rush to treat patients. Instead, he began chemical and physiological experiments to establish the parameters of the new technique and developed an inhaler that took into account the relation between temperature and the "strength" of the vapor. In just 6 months, Snow described the different degrees of anesthesia that marked ether's sequential suspension of consciousness and volition. Snow mastered the intricacies of ether, whereas other physicians struggled. Ether's irritant qualities made it difficult to breathe, and ill-designed inhalers, which gave too little of the vapor, stimulated the patients rather than subduing them.

In the United States, ether inhalation became a part of social entertainment, and there were "ether parties." On March 30, 1842, Crawford Long (1815–1844), a physician who had used ether for amusement when he was a student at the University of Pennsylvania successfully employed it for the painless surgical removal of a tumor of the neck after the patient (a student) inhaled the ether. A second patient, a child who required amputation of two severely burned fingers, was given ether, but when he awoke prematurely and was in great pain, he had to be tied down to complete the operation. Under popular pressure, Long was forced to abandon his use of ether.

In the late 1840s, American dentists, especially William T. G. Morton, who had previously employed ether for tooth extractions, promoted its use in surgery. The first public demonstration of its successful use took place in the surgical amphitheater (now named the Ether Dome) at the Massachusetts General Hospital on October 16, 1846. A large vascular tumor of the neck was removed painlessly by the surgeon John Collins Warren as Morton administered ether by using a special but simple device: a sponge soaked in ether placed inside a glass vessel with a spout that was placed in the patient's mouth. That day is now called Ether Day. Morton concealed the chemical nature of the ether by adding fragrances and called his drug "Letheon," from the Greek mythology, in which a drink from the river Lethe could expunge all painful memories. He applied for and received a U.S. patent for Letheon, but the patent was later denied. He also applied for a monetary reward ($100,000) for his discovery from the federal government, and this too was denied. Charles Jackson, a Boston physician who had advised Morton to use ether, also claimed to have played a large part in its discovery; they both pressed for credit all the way to Congress, but on August 28, 1852, the U.S. Senate voted 28–17 to deny Morton's claim of priority in the discovery of the anesthetic properties of ether, and Morton was forced to reveal that Letheon was nothing more than ether. Morton was crushed, and in 1853, he was expelled from the American Medical Society and disowned by fellow dentists for his unseemly and greedy conduct. He then moved to a cottage near Boston, which he named "Etherton," and died on July 15, 1868.

Jackson claimed that as early as 1845, he had used ether to banish pain and that he had used it for the extraction of teeth. However, because very few in the profession rallied to support his claim, he turned to an old-fashioned anesthetic, alcohol. In 1873, he was involuntarily committed to McLean Asylum, a mental institution, where he remained until his death on August 28, 1880. For Jackson's contributions to the discovery of ether anesthesia, Massachusetts General Hospital underwrote all of the expenses during Jackson's stay at MacLean.

Wells, Long, Jackson, and Morton, but especially Morton, deserve some measure of credit for the discovery of etherization. Despite Morton's greed, he probably should receive most of the credit for his public demonstration of general anesthesia. It was through all of their efforts that dentists and surgeons soon began to catch on to the benefits of ether anesthesia, and on December 21, 1847, Liston (the surgical speed artist) performed the first painless major operation (a leg amputation) by using ether.

Chloroform

Chloroform is a highly volatile substance with an agreeable taste and smell. It was discovered simultaneously in 1831 by the chemists Simon Guthrie in New York, M. Soubeiran in France, and Justus von Liebig in Germany. Given the name chloroform by M. Jean-Baptiste Dumas, none of the chemists expected it to be of any practical use, until James Simpson (1811–1870) tried to inhale its fumes. Simpson, an Edinburgh obstetrician, had experimented with chloroform at a "chloroform frolic party," attempting to produce the same effects as N_2O, but when all of the guests fell asleep, he realized the potential of the substance as an anesthetic. On recovering consciousness, Simpson said, "This is stronger than ether." On November 4, 1847, Simpson used chloroform to relieve the pain of labor during childbirth. Other successes in both Britain and the United States soon followed Simpson's public announcement in 1848. However, the Scottish Church vehemently opposed the use of this anesthetic to relieve pain during childbirth, because the Bible says, "in sorrow thou shalt bring forth children." Simpson responded in kind with: "And the Lord caused a deep sleep to fall upon Adam; and he slept; and He took one of his ribs, and closed up the flesh instead thereof."

The suffragettes, however, knew that the discrimination against painless childbirth came from the male-dominated medical establishment and the clergy and promoted the use of chloroform and ether. In 1850, Simpson used ether to deliver Queen Victoria's seventh child, Arthur. The London physician John Snow used chloroform during the birth of Queen Victoria's eighth child, Leopold, in 1853, and Queen Victoria described it as "the blessed chloroform." In addition, when chloroform was used at the birth of her ninth child, Beatrice, in 1857, Victoria said, "We are having a child and we are having chloroform." Snow anesthetized 77 obstetric patients with chloroform, and in 1858, he wrote a treatise, "Chloroform and Other Anesthetics," in which he wrote: "The most important discovery that has been made in the practice of medicine since the introduction of vaccination is undoubtedly the power of making persons perfectly insensible to the most painful surgical operations, by the inhalation of ether, chloroform and other agents of the same kind." He also described 50 deaths during chloroform administration and recommended means for prevention. Simpson was knighted for his development of chloroform anesthesia, but not Snow (who had died in 1858). Because of its quick action and ease of administration, chloroform revolutionized battlefield surgery. During the American Civil War, the production of ether and chloroform by Squibb and Company (established in 1858) enabled it to reap substantial profits.

Cocaine

For over 1000 years, the indigenous people of Peru chewed the leaves of the coca plant (*Erythroxylum coca*); indeed, the remains of coca leaves have been found with ancient mummies, and pottery from that time period depicts humans with bulged cheeks, indicating the presence of something that they were chewing. There is also evidence that the Incas used a mixture of coca leaves and saliva as an anesthetic for the performance of surgery (trephination). In 1539, Vicente de Valverde, who accompanied Francisco Pizarro during the conquest of the Incas, wrote: "coca ... the leaf of a small tree ... is the one thing that the Indians are ne'er without in their mouths, that they say sustains them and gives them refreshment, so that even under the sun they feel not the heat and it is worth its weight in gold in these parts." Initially, the Spaniards wanted to ban coca as an evil agent of the devil; however, when they realized that the Inca workers were less productive in the field and in the gold mines without the stimulant effects, they decided to legalize and tax it, taking 10% of the value of the crop.

In 1609, Padre Blas Valera wrote: "Coca protects the body from many ailments, and our doctors use it in powdered form to reduce the swelling of wounds, to strengthen broken bones, to expel cold from the body or prevent it from entering, and to cure rotten wounds or sores that are full of maggots." In 1859, Carl Scherzer managed to import a large quantity of coca leaves into Germany and gave these to Professor Friedrich Wohler at the University of Gottingen. Wohler, in good academic fashion, passed these leaves of coca on to his graduate student Albert Niemann, suggesting that isolation of some of its chemical constituents might serve as a suitable topic for his doctoral studies. In 1860, Niemann duly obliged his professor and in his doctoral thesis described in great detail every step he took in the isolation of a white powder that he named cocaine. He described cocaine as "colorless transparent prisms that have an alkaline reaction, a bitter taste and promote the flow of saliva," and when Niemann placed it on his tongue, it produced numbness and a sense of cold.

By the 1880s, cocaine became the rage among those in the medical profession. In 1884, Sigmund Freud's "Cocaine Papers" described in considerable detail the medical experiments that he and others had conducted with cocaine. Freud's interest in cocaine began after he had read an 1859 paper (on the hygienic and medical values of coca) by the prominent Italian neurologist Paolo Mantegazza that extolled the virtues of cocaine on the basis of self-experimentation. Mantegazza took small doses of cocaine and claimed that his digestion and pulse were promoted and that, with larger doses, he experienced flashes of light, headaches, and increased mental and physical vigor. On April 21, 1883, Freud wrote to his fiancée that he was "toying … with and a hope. … It is a therapeutic." As a scientist, Freud was not satisfied with merely reviewing the reports on the experiments that others had conducted with cocaine. Freud was a physician who trained himself as an experimental subject. Nine days after writing the letter, Freud took cocaine for the first time. In these self-experiments, Freud established the appropriate dosage and time course of the drug's action on muscular reactions and psychic reaction time. Freud recommended coca or cocaine for a variety of illnesses. He based his claims on "some dozen" experiments that he had conducted on himself and the additional ones on others. Freud and his colleagues noted that cocaine numbed their tongues and cheeks. Although others had described the same sensation earlier, none had the foresight to see its application to surgery, until Freud's friend, an ophthalmologist, Carl Koller (1857–1944), applied it.

In the 1800s, eye surgery was all but impossible, because there was no way to prevent eye movement during a surgical operation. Although general anesthesia eliminated pain, a patient under the influence of ether or N_2O could not cooperate with the surgeon's directions to gaze left or right. Further, N_2O and ether often caused retching, vomiting, and agitation, which could raise the pressure in the eye and destroy the skilled work of the surgeon, because healing of the eye required rest and immobilization. After learning of the anesthetic properties of cocaine, Koller prepared a solution of cocaine and dropped it into the eye of a frog. The initial irritation caused the frog to blink for up to 30 seconds, but then, the blinking stopped and the frog's eye stared straight ahead. When Koller pricked and scratched the frog's eye with a needle, the frog remained still. Koller repeated the experiments with guinea pigs, rabbits, and dogs, and finally on himself. When he touched the cornea with a pin, he said, "I can't feel a thing." Koller tried the same procedure with his patients and it worked. He found a local anesthetic that would serve two purposes: a painkiller for people with eye conditions and an anesthetic for eye surgery.

Following these discoveries of the anesthetic properties of cocaine, other physicians wondered whether they could also use it in minor surgery. In 1884, using the hypodermic syringe introduced by Alexander Wood in 1853 in Scotland, two surgeons, Richard J. Hall and William S. Halsted, at New York's Roosevelt Hospital, injected cocaine through the skin into or near the sensory nerves to block pain sensations. Using this technique, they were able to carry out minor operations with local anesthesia on conscious patients. Hall and Halsted used cocaine as a local anesthetic for thousands of operations in New York. However, before doing so, they experimented on themselves. Hall and Halsted repeatedly tested the effects of different doses of cocaine, injecting it at various depths into their arms and legs; they used both strong and weak solutions of cocaine. In the September 12, 1885 issue of the *New York Medical Journal* Halsted described his findings; however, the article was disjointed and overwrought, and it is suspected that he had written the article under the influence of cocaine. Although Hall, Halsted, and two other assistants made major contributions to surgical anesthesia, they paid a heavy price: they became accidental cocaine addicts. They all died, without recovering from the habit.

Halsted also injected cocaine into the lower jaw of a patient and then extracted a tooth, with no pain or sensation felt by the patient. It was not long thereafter that cocaine was widely adopted as a local anesthetic by dentists. The feelings of euphoria were not lost on those dentists administering cocaine, and soon, many dentists developed a dependency on it. Once this was observed on a large scale, chemists quickly sought to develop an alternative to cocaine for use as a local anesthetic. In 1905, Alfred Einhorn (1856–1917), a chemist at Hoechst, synthesized a new local anesthetic, procaine. Procaine (proprietary name, Novacain) had all the desired effects as a local anesthetic, but it neither possessed the addictive potential, nor had the other negative side effects of cocaine such as elevations in heart rate and blood pressure, as well as provoking of potentially lethal dysrhythmias, for example, ventricular fibrillation. It quickly replaced cocaine as the local anesthetic of choice in dentistry. Procaine, like cocaine, is an amino ester that works by diffusing through the lipid-rich nerve membrane and then blocking sodium ion channels, thus producing a nondepolarizing nerve block. Anesthesia is terminated when the drug diffuses out of the sodium ion channels. It is then redistributed to other areas in the body via the bloodstream, and there, it is metabolized via plasma pseudocholinesterase. Novocain had its own drawbacks, so a new generation of local anesthetics was developed. In 1943, in Sweden, Nils Lofgren synthesized a new class of local anesthetic, developing lidocaine, the first amino amide. Marketed in 1948 under the proprietary name Xylocaine, it quickly became a favorite of the dental profession, replacing procaine as the "gold standard." Lidocaine's onset of action was measurably faster (3–5 minutes); its duration of anesthesia was longer and more profound; and it provided more consistently reliable anesthesia than did the esters.

Coca-Cola and Cocaine

When cocaine and ethyl alcohol are ingested together, the liver converts them into a unique drug cocaethylene; cocaethylene works like cocaine by inhibiting the reuptake of serotonin, norepinephrine, and dopamine to result in greater concentrations of these three neurotransmitters in the brain, but it has a longer duration of action than cocaine, and in most users, it produces euphoria. In 1863, a Parisian chemist Angelo Mariani combined cocaine with claret. The combination of good Bordeaux and cocaine sold extremely well. Mariani claimed it to be a stimulant for the fatigued and overworked body and brain. "It prevents malaria, influenza and wasting diseases." His Vin Mariani, containing 11% alcohol and 6.5 mg of cocaine per ounce, became extremely popular and was endorsed by Pope Leo XIII and the Chief Rabbi of France. Jules Verne, Alexander Dumas, and Arthur Conan Doyle were among literary figures said to have used it. Eventually, Mariani marketed a number of other cocaine products, in addition to his wine, including tea, lozenges, and cigarettes. In 1884, seeing this commercial success, John Styth Pemberton, a drugstore owner in Columbus, Georgia—himself a morphine addict following an injury in the Civil War—set out to make his own version of Vin Mariani. He called it Pemberton's French Wine Coca and marketed it as a panacea. Among its many fantastic claims, he said it was "a most wonderful invigorator of sexual organs" and a "brain tonic and intellectual beverage." However, as Pemberton's business started to take off, a prohibition was passed in his county in Georgia (a local one that predated the Eighteenth Amendment by 34 years). Soon, French Wine Coca became illegal—because of the alcohol, not the cocaine.

However, Pemberton remained a step ahead. He replaced the wine in the formula with sugar syrup and added an extract of kola nuts (a natural source of caffeine). His new product, Coca-Cola, a combination of caffeine and cocaine, debuted in 1886. It was marketed and promoted as a healthier alternative to the more dangerous alcoholic-containing beverage. However, by 1903, as public sentiment and the press turned against cocaine, it was eliminated from the drink. The Coca-Cola we know today still contains coca, but the ecgonine alkaloid—the psychoactive part—is extracted from it. Perfecting this extraction process took until 1929. A company called Stepan Company, the only legal commercial importer of coca leaves, now does the extraction at a heavily guarded plant in Maywood, New Jersey. The Stepan Company imports approximately 100 metric tons of dried coca leaves each year. The cocaine-free leaves are sold to the Coca-Cola Company, whereas the extracted material is sold to Mallinckrodt Inc., the only company in the United States licensed to purify the product into cocaine hydrochloride for medicinal purposes.

In the United States, the manufacture, importation, possession, and distribution of cocaine are regulated by the Controlled Substances Act of 1970. In 2013, 419 kg of cocaine was produced legally. The illegal market for the drug is estimated at $100–500 billion per year. The most common methods of ingestion of recreational cocaine hydrochloride are nasal inhalation ("snorting" or "sniffing" or "blowing") and the smoking of "crack" produced by the neutralization of cocaine hydrochloride with a solution of baking soda and water, yielding a soft mass that hardens into a rocklike state on drying. (The origin of the term "crack" comes from the cracking sound when the cocaine is heated past the point of vaporization.) Another form of cocaine that can be smoked is the "free base," produced by dissolving cocaine hydrochloride in ammoniated water and with ether added to dissolve the cocaine base. The "free base" is then extracted from the ether through evaporation. If removed before extraction is complete, there may be residual ether that is highly flammable and may cause facial burns when the freebase is smoked. Cocaine may also be injected. It is estimated that there are over 6 million Americans over the age of 12 years who have abused cocaine in some form, and in 2013, cocaine use resulted in 4300 deaths.

Morphine

Of the more than 100 different kinds of poppies, only one, *Papaver somniferum*, produces the substance opium in sufficient quantities to make it an important medicine. The milky juice from the seedpods of *P. somniferum*, containing some 30 or so alkaloids, some of which relieve pain, relax muscle spasms, reduce fevers, and induce sleep, has been used since early humans domesticated it from the wild strain *P. setigerum*. The most important of the opium alkaloids are morphine and codeine.

The Egyptians, Greeks, and Romans, all enjoyed the opium poppy. In ancient Egypt, it was placed in the pharaohs tombs. In the Greek classic, *The Odyssey*, it was used to ease the symptoms of depression. The Greek physician, Galen, used opium to treat a variety of illnesses, including vertigo, epilepsy, headaches, melancholia, deafness, asthma, jaundice, leprosy, and malaria fevers. Although it is unlikely that treatment with opium actually produced any cures, it did put the patient in a state of mind so that he was no longer concerned with his ailment. By the eighth century, opium use had spread from the Fertile Crescent to India, China, and the Arabian Peninsula. In 1650, Christopher Wren and Robert Boyle noted that intravenous opium (using the quill of a pen) induced sleep in a dog. In 1700, John Jones wrote a treatise on opium (*Mysteries of Opium Revealed*), detailing the acute and chronic effects of opium. At the time of the writing, opium was usually taken in Europe as laudanum (an alcoholic preparation of opium), often mixed with wine and a variety of other ingredients to disguise its bitter taste. In the eighteenth century, opium use in Western Europe increased, largely influenced by Dr. John Brown's *Elements of Medicine*, which recommended the stimulatory powers of opium for increasing the body's level of excitation, thereby countering the body's lack of excitation, which he and others believed characterized many diseases. Indeed, Brown regularly took opium for his gout.

By the 1800s, opium was the painkiller of choice for physicians in Europe, but its effects were unpredictable largely because each batch differed in its potency. All this changed after the pharmacist-chemist Friedrich Serturner (1783–1841) was able to isolate the sleep-making principle from raw opium. In 1799, Serturner began his apprenticeship in pharmacy in the north-central German town of Paderborn. He completed his studies in 4 years, and on August 2, 1803, he passed the examination for pharmacy assistants. Serturner, aware of the complaints by physicians over the unpredictability of opium's analgesic properties, theorized that the problem would not be remedied until the active ingredient from opium had been isolated. Once this would be accomplished, he believed, it would be possible to produce predictable and reliable doses. In his quest to find the active substance, Serturner was one of the first to apply the basic techniques of chemical analysis to pharmacology. In 1805, working evenings and with very old equipment, he patiently went through a series of experimental steps—57 in all—and eventually found that when opium was treated with ammoniated water, it yielded a yellow-white crystalline powder. Confounding conventional wisdom of the time, the narcotic substance he isolated was an alkaloid, the first such derived from a plant source. Serturner named the pain-killing substance he isolated "morphine" after the Greek god of sleep, Morpheus. Once he had isolated morphine, Serturner began testing it on animals. He put crystals of pure morphine in food for mice in the cellar and in the food for some stray dogs. In both cases, the animals fell asleep, and ultimately, they died. Undaunted, he reduced the dosage, and because he felt that experiments with animals did not give exact results, he tried it on himself and three of his young friends, each 17 years old. Using increasing amounts of morphine dissolved in alcohol, he found that with the lowest dose, they became feverish, and with increased amounts, the fever became more severe and they became nauseous and dizzy. Finally, they experienced pain and palpitations and then they fell asleep. The following day, they had a lack of appetite and headaches and stomachaches that persisted for several days. "They had swallowed about 10 times the amount of morphine now recommended." Serturner concluded, "the principal effects of opium are dependent on pure morphine and it is to be wished that qualified physicians might concern themselves with this matter because opium is one of our most effective drugs."

Although Serturner published his findings in 1805, it received little attention. Six years later, he again tried to draw attention to his work but was again unsuccessful. However, his

1817 publication did attract the attention of the French physician Joseph Guy-Lussac, who recognized its importance. In 1818, the French physician Francois Magendie published a paper on morphine's pain-relieving and sleep-inducing qualities, which attracted the attention of pharmaceutical companies, including George Merck, who turned his family's seventeenth-century pharmacy in Darmstadt, Germany, into a major supplier of standardized doses of the drug.

Heroin

Morphine was widely used for pain relief during the American Civil War (1861–1865) and during the Franco-Prussian War (1871–1872). When used in combination with the hypodermic syringe (invented in 1853), precisely measured amounts of the purified drug could be administered and with increased potency. After its heavy use during the American Civil War, physicians began to realize that morphine was more addictive than opium. In 1874, while seeking a less addictive alternative to morphine, the chemist Charles Alder Wright, working at St. Mary's Hospital Medical School in London, synthesized a diacetyl derivative of morphine by boiling anhydrous morphine with acetic anhydride and acetic acid over a stove for several hours. After running a few experiments with it on animals, though, he abandoned his work on the drug. Twenty-three years later, Felix Hoffman, working at Bayer, in Germany, took interest in acetylating morphine (as he had done with salicylic acid) and managed to independently synthesize diacetylmorphine. Diacetylmorphine was found to be significantly more potent than morphine, and so, Heinrich Dreser, head of the pharmacological laboratory at Bayer, decided that they should move forward with it rather than with another drug that they had recently created (Aspirin). Dreser was apparently well aware that Wright had synthesized diacetymorphine years before, but despite this, he claimed that Bayer's synthesis was an original product, and by early 1898, Bayer began animal testing, primarily on rabbits and frogs. Next, they moved on to testing it on people, primarily workers at Bayer, including Heinrich Dreser himself. After successful trials, Bayer named the drug heroin from the German adjective "heroisch" (meaning heroic), since nineteenth-century doctors called such a medicine powerful.

In the late 1800s, painful respiratory diseases such as pneumonia and tuberculosis were the leading causes of death, and in the days before antibiotics, the only remedy that physicians prescribed were narcotics to alleviate patient suffering. In 1898, Dreser presented heroin to the Congress of German Naturalists and Physicians as more or less a miracle drug that was "10 times" more effective than codeine as a cough, chest, and lung medicine and worked even better than morphine as a painkiller. He also stated that it had almost no toxic effects, including being completely nonaddictive. Dreser particularly pushed the potent and faster-acting heroin as the drug of choice for treating asthma, bronchitis, and tuberculosis. In early 1898, G. Strube of the Medical University Clinic in Berlin tested oral doses of 5 and 10 mg of heroin in 50 tuberculosis patients to relieve coughing and to induce sleep. He did not notice any unpleasant reactions, and most patients continued to take the drug after he no longer prescribed it. Because heroin worked well as a sedative and respiration depressor, it did indeed work extremely well as a type of cough medicine and allowed people affected by debilitating coughs to finally be able to get some proper rest, free from coughing fits. Further, because it was marketed as nonaddictive, unlike morphine or codeine, it was initially seen as a major medical breakthrough and was quickly adopted by the medical profession in Europe and the United States.

Just one year after its release, heroin became a worldwide hit, despite it not actually being marketed directly to the public but rather simply to physicians. Heroin was soon sold in a variety of forms: mixed in cough syrup; made into tablets; mixed in a glycerin solutions as an elixir; and put into water-soluble heroin salts, among others. At the end of this first year, it was popularly sold in over 23 countries, with Bayer producing around 1 ton of it in that year. However, it quickly became apparent that Bayer's claims of the drug not being addictive were completely false, with reports popping up within months of its widespread release. Despite this, it continued to sell well in the medical field. The United States was the first country in which heroin addiction became a serious problem. By the late nineteenth century, countries such as Britain and Germany had enacted pharmacy laws to control dangerous drugs, but under the U.S. Constitution, individual states were responsible for medical regulation. Late in the century, some state laws required morphine or cocaine to be prescribed by physicians, but drugs could still be obtained from bordering states with laxer regulations. Moreover, this era was the peak of a craze for over-the-counter "patent" medicines that were still permitted to contain these drugs. At the turn of the century, it is believed that over a quarter of a million Americans (from a population of 76 million) were addicted to opium, morphine, or cocaine.

After years of resistance, American patent medicine manufacturers were required by the Federal Pure Food and Drug Act of 1906 to accurately label the contents of their products. These included "soothing syrups" for crying babies and "cures" for chronic illnesses such as consumption or even drug addiction, which previously had not declared (and sometimes denied) their content of opium or cocaine. By this time, consumers were becoming fearful of addictive drugs so the newly labeled patent medicines either declined in popularity or their addictive ingredients were removed. (The preeminent survival from this era was a beverage from Atlanta called "Coca-Cola.") In 1914, President Woodrow Wilson signed the Harrison Narcotic Act, which exploited the federal government's power to tax as a mechanism, finally enabling federal regulation of medical transactions in opium derivatives or cocaine. In addition, in 1924, the U.S. Congress banned all domestic manufacture of heroin. Now, under federal law and the laws of all 50 states, heroin possession is a crime.

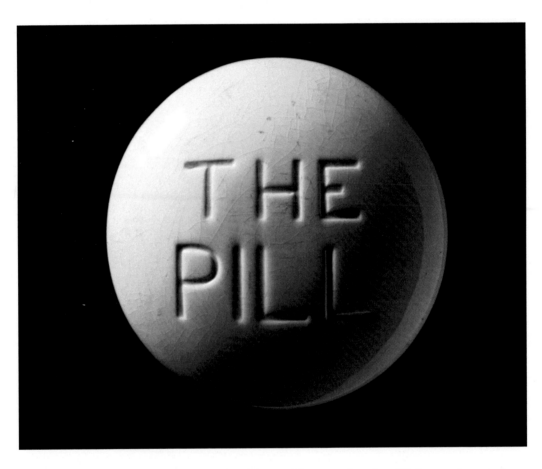

The Contraceptive Pill. (Courtesy of the Science Museum, London, Wellcome Images.)

Chapter

4

The Pill

In 2010, a very special 50th birthday party was being held at the Hotel Pierre in New York City. A couple of hundred bejeweled women were in attendance. The reception room was painted robin's egg blue and the bar was festooned with silver stars. In the middle of the hotel's lavish ballroom, on a tall pedestal, was perched an enormous cake with bold letters: "ONE SMALL PILL, ONE GIANT LEAP FOR WOMANKIND, ONE MONUMENTAL MOMENT IN HISTORY." The celebrated "birthday girl" was the contraceptive birth control pill that had been approved in May 1960. The "pill," the most popular form of birth control in the United States, was due (in part) to the efforts of an academic chemist Carl Djerassi (1924–2015) who produced synthetic progesterone (called progestin) that could be taken orally to suppress ovulation, thus enabling millions of women to chart their own reproductive destiny.

Djerassi's success was not a solo effort; it followed the work of others on the chemical control of reproduction. As early as 1919, Ludwig Haberlandt of the University of Innsbruck was able to achieve "hormone temporary sterilization" when he implanted the ovaries of a pregnant rabbit into another rabbit, which, despite frequent coitus, remained infertile. This was proof of the principle of the possibility of hormonal control of contraception; however, its application was impractical for several reasons: it required surgery and the material present in the gland would secrete a mixture of materials some of which might be toxic. Haberlandt recognized that it was the corpus luteum—the yellow body left in the ovary after egg release—that carried the sterilization factor. He contacted several pharmaceutical companies to enlist their assistance in purifying the active, nontoxic material from the corpus luteum and by 1931 had in hand a sterilizing preparation; he named it "Infecundin." By oral administration, it worked in mice, albeit poorly. It was planned to be used for clinical experiments by Gedeon Richter Ltd, but Haberland's untimely death at age 47 stopped this project in 1932.

In the early 1930s, pure progesterone—the corpus luteum hormone—was isolated in laboratories in Germany, the United States, and Switzerland; its chemical structure had been determined, and it could be synthesized from a plant sterol stigmasterol. In 1937, A.W. Makepeace and E. W. Dempsey showed that injections of pure progesterone were able to inhibit ovulation in rabbits and guinea pigs. Although progesterone was a sterilizing preparation, it had to be administered by injection because it lacked the chemical features that would enable it to withstand the enzymes and acids of the digestive tract after oral administration. Clearly, a different type of progesteronal preparation would be necessary to serve as an effective oral contraceptive.

From Yam to Progestin

Preparation of a novel substance with progesteronal properties came from two unlikely sources: a brilliant, single-minded, perseverant maverick, Russell Marker (1902–1995), and a Mexican yam. Marker, a Maryland farm boy, disregarded fatherly advice and instead of majoring in agriculture enrolled at the University of Maryland as a chemistry major receiving a BS in chemistry (1923) and a master's degree in physical chemistry (1924). Marker produced enough data to write a doctoral thesis in 1 year of work, and all that stood between him and a PhD was two required physical chemistry courses. But Marker refused to take physical chemistry inasmuch as he already had a master's degree in physical chemistry. The university, however, refused to modify its graduation requirements. Marker's own thesis advisor threatened him with a dead-end career of a "urine analyst" if he did not complete his course work, but Marker stubbornly refused and left the university without the completion of the doctoral degree. Marker's interest in hydrocarbon research led him to the Ethyl Corporation where he developed an octane rating system for gasoline that is still in use today. Following this, he worked as an organic chemist at the Rockefeller Institute for Medical Research where, over a 6-year period, he worked with the nucleic acid chemist P. A. Levene. Marker wanted to shift his work into steroid chemistry; however, Levene was not interested, so in 1934, Marker left Rockefeller for a Parke-Davis funded position at Pennsylvania State College (now University). At the time, the only known way to isolate progesterone was by a laborious and inefficient process. Yet, by 1937, Marker had extracted enough pregnanediol from the urine of pregnant cows and mares to produce 35 g of progesterone, which was worth $1000 a gram at the time. Realizing that a reliable and abundant source of progesterone would be a gold mine, Marker began to look for plant steroids that could serve as precursors for progesterone. Initially, he theorized that it should be possible to convert the steroid in the root of the sarsaparilla plant, saraspogenin, into progesterone. He succeeded in removing the complex side chain by a series of oxidative reactions (Marker degradation) and was able to produce progesterone in low yield from saraspogenin. However, saraspogenin itself was very expensive ($80 per gram, seven times more costly than gold), so he sought a better and less expensive source. Marker began collecting plants from the American Southwest examining more than 400 species of yam-like plants and isolating a dozen new sapogenins; however, none fit the bill. Then in 1941, Marker was reading a textbook of botany and saw a picture of a yam (*Dioscorea*) that grew in Vera Cruz, Mexico, called "cabeza de negro" with a huge root often weighing 200 pounds or more (Figure 4.1). Marker went to Vera Cruz, enlisted the help of a local, and was able to harvest two roots. One was confiscated at the border, but the

FIGURE 4.1 Russell Marker with the yam "cabeza de negro."

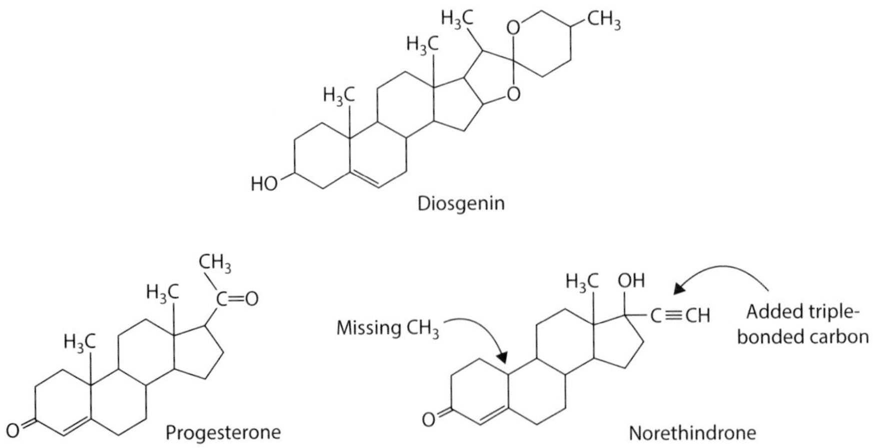

FIGURE 4.2 A comparison of the structure of diosgenin with progesterone and norethindrone.

larger 50-pound root was *passed* on the strength of a bribe. The root, loaded with the soap-like saponin, diosgenin, was employed by the natives for doing laundry or to daze and kill fish, but Marker used it in another way. In a five-step Marker degradation process, he was able to extract the diosgenin from 10 tons of dried "cabeza de negro" into 3 kilos of progesterone (worth $80 a gram) (Figure 4.2). In 1944, unable to convince any American pharmaceutical company of the commercial potential of diosgenin, Marker joined Emeric Somlo and Frederico Lehmann, two immigrant owners of a small pharmaceutical company in Mexico, Laboratorio Hormona, to form a new company named Syntex (from Synthesis and Mexico). Within a year, Syntex was selling progesterone for $50 a gram. After 2 years, however, there was a rancorous dispute between Marker and his partners over profits and their distribution; Marker severed all ties with Syntex and left the company. Marker then formed a new company, Botanica-mex, and over several months produced progesterone. Marker had great difficulty. A man collecting roots was killed, a guard was shot through his leg, and one of the women was choked and beaten. Because of the harassment the workers experienced at Botanica-mex, production was discontinued in 1946, and the assets of the company were sold to Gedeon Richter Ltd. By 1949, Marker retired from chemistry and dedicated himself for the next 30 years to commissioning Mexican-made replicas of antique European silver works.

Earlier, Marker had published his methods in the *Journal of the American Chemical Society* (*JACS*), and because no one had taken out a patent in Mexico, the preparation of progesterone from diosgenin was up for grabs in that country. To continue the manufacture of progesterone, Somlo and Lehmann recruited another chemist, Dr. George Rosenkranz, a Hungarian refugee then living in Havana. At Syntex, Rosenkranz, who had received his PhD in Switzerland under the direction of the steroid chemist and Nobel laureate Leopold Ruzica, reinstituted the large-scale production of progesterone as well as testosterone from diosgenin. By the late 1940s, Syntex was the bulk supplier of these hormones.

In the spring of 1949, the then 25-year-old Carl Djerassi (then working as a research chemist at CIBA) received an unsolicited employment offer from the then 30-year-old Rosenkranz at Syntex. At first, Djerassi was disappointed when he was shown a rather crude laboratory, but he was then elated by Rosenkranz's promise of a large number of laboratory assistants, substantial research autonomy, and the guarantee that any scientific discoveries could be published in chemical journals without patent attorneys making the decision to publish or not to publish. This extraordinary approach by a pharmaceutical company was what attracted Djerassi to Syntex who was determined to publish his work in reputable scientific journals and eventually to become an academician.

Carl Djerassi (pronounced jer-AH-see) was born in Vienna on October 29, 1923, to assimilated Jewish physician parents, Samuel and Alice Djerassi. Carl was a brilliant student, attended schools in Vienna and summered in Sofia, Bulgaria, where his father specialized in treating venereal disease with salvarsan before penicillin became available. In 1938, as World War II engulfed Europe, and Austria was annexed by Nazi Germany with the rounding up of Jews and Communists, Carl and his mother decided to emigrate. In 1939, penniless, the two arrived in New York City. Carl enrolled in Newark Junior College. After writing a letter to Eleanor Roosevelt, who passed it on to the Institute of International Education, he received a scholarship to Tarkio College in Missouri. It was a little help that made a huge difference. By the fall of 1941, he entered Kenyon College in Ohio and was determined to get a bachelor's degree within a year. He majored in chemistry and graduated summa cum laude. His first job was as a junior chemist at the Swiss pharmaceutical company, CIBA, in New Jersey. At CIBA, he was part of a team credited with the discovery of an antihistamine (pyribenzamine), had published a number of papers on steroids, and started to take graduate courses in chemistry at NYU and Brooklyn Polytechnic. After a year at CIBA, he received a Wisconsin Alumni Research Foundation Fellowship for graduate study at the University of Wisconsin. At Wisconsin, his PhD advisor was Alfred L. Wilde who along with Werner Bachman and Wayne Cole had been able to achieve the total synthesis of equilenin, a simple estrogenic hormone found in horse urine. Djerassi's doctoral thesis was a compromise between Wilde's interests in total synthesis of steroids and his own interests in the transformation of intact steroids, specifically testosterone into estrogen. By the fall of 1945, Djerassi, now an American citizen, returned to CIBA (which had supported him at Wisconsin with a supplementary stipend) to work on antihistamines and other medicinal compounds.

In 1949, Philip Hench and Edward Kendall at the Mayo Clinic reported that rheumatoid arthritis could be "cured" with cortisone. There was much ballyhoo with newspapers featuring stories of crippled arthritics dancing in the streets after therapy. The medical community clearly recognized the importance of treating inflammatory diseases with cortisone; however, because of its expense, therapeutic use was limited. Indeed, until 1951, the only source of cortisone was a very complex process requiring 36 chemical transformations starting with ox bile that had been pioneered by Merck. Because CIBA allowed its chemists to spend 20% of their time on independent research, Djerassi wanted to carry out research on cortisone; however, permission was not granted because that work was being carried out in Switzerland. The restless Djerassi wanted a new challenge. A chemist friend from Schering proposed that he consider the offer from Syntex. Arriving at Syntex in Mexico City in early 1951, his first project was to synthesize cortisone from diosgenin, and by August, he and his team of young Mexican assistants had more than four communications to the editor of *JACS* dealing with different synthetic approaches to cortisone. *Life* magazine recognized the feat and featured a huge photograph of the team with the headline "Cortisone from Giant Yam, Beat Out Competition from Harvard and Merck."

In July 1951, Rosenkranz received a phone call from Upjohn requesting that Syntex supply 10 tons of progesterone. The quantity was unheard of, and Syntex concluded that Upjohn was using the progesterone as an intermediate rather than as a hormone on its own. Indeed, it was only a few months later that the Syntex achievement of cortisone synthesis from diosgenin was overshadowed by the discovery at Upjohn that in a single step and in a few hours a microbe did with its own enzymes what Syntex had accomplished more laboriously through many chemical transformations. Now, Djerassi turned his attention to the synthesis of synthetic progesterone (a progestin) that could be used to treat certain kinds of infertility and menstrual disorders and could be taken orally. The key would be to replace a single carbon atom (number 19) in the steroid ring of estrogen with a hydrogen atom.

During World War II, Hans H. Inhoffen at Schering in Berlin had been able to achieve this feat, and Syntex used a similar process to produce small quantities of the estrogens: estrone and estradiol. Djerassi suggested to Rosenkranz that they try to make estrogens from testosterone using the Inhoffen process, and using this as a base, various chemical methods were developed to prepare pure crystalline 19-nor progesterone (which lacks the carbon atom 19). Because Syntex lacked facilities for biological testing, the compound was sent to Endocrine Laboratories in Wisconsin for assay. In rabbits, it was found to be four–eight times as active as natural progesterone as a sterilizing agent. Turning to another find made by Inhoffen in 1939, Syntex chemists added an acetylene group at position 17 in the testosterone molecule and found this modification made the product insensitive to degradation in the digestive tract; more importantly, instead of retaining male hormone properties, it had progesteronal activity. Putting all this together Rosenkranz, Luis Miramonte, a young Mexican chemist doing his bachelor's degree at Syntex, and Djerassi synthesized (on October 15, 1951) the 19-nor analog, 19-nor-17α-ethynyltestosterone, which they named norethindrone (Figure 4.2). Submitted to the Endocrine Laboratories in Wisconsin for biological evaluation, it was found to be more active as an oral progesteronal hormone than any other steroid at the time. The patent for norethindrone was filed on November 22, 1951, and the full article describing all experimental details appeared in the *JACS* in 1954.

Development of the Contraceptive Pill

At the time of the filing for the patent for norethindrone, the hormones involved in regulating the 28-day ovarian and uterine cycle of women were well known: eggs begin to ripen in the ovary under the influence of two hormones from the pituitary gland (follicle stimulating hormone [FSH] and luteinizing hormone [LH]); release of the egg (usually on day 14) is triggered by a brief but accelerated release of pituitary LH, and after release of the egg from the ovary, the remnant called the corpus luteum begins to secrete estrogen and progesterone. The estrogen and progesterone circulate in the blood, act on the endometrial lining of the uterus to thicken it, and prepare it for implantation of the fertilized egg. The progesterone and estrogen arrive at the pituitary and act on it to suppress the production of FSH and LH, so that no new eggs develop. If the egg is not fertilized, the corpus luteum becomes inactive (usually on day 22 or 23 of the cycle), the levels of progesterone and estrogen decline, and the endometrial lining of the uterus breaks down (menstruation). The decreased production of progesterone and estrogen causes removal of the block on FSH production by the pituitary, and new FSH production initiates new ovarian and menstrual cycles. Understanding how hormones influence the ovarian cycle provided the background for developing a method for suppressing ovulation, the oral contraceptive pill.

Margaret Sanger (1879–1960) was the sixth child of a poor working class Irish Catholic family living in New York. She witnessed the death of her own mother after giving birth to what she felt were too many children. During the 1920s and 1930s, Sanger worked as a visiting nurse in the slums of New York City and saw women in the tenements who were wrung out after having 8, 9, and 10 children with no idea how to stop it other than abortion. She observed women who died from botched abortions, as well as the many young women who died in labor. In 1914, she coined the term "birth control" and dared to use the phrase in a publication. For this crime and others, she was indicted for violations of the Comstock Law passed by Congress in 1873 that specifically listed contraceptives as obscene material and outlawed dissemination of them via the postal services or interstate commerce. Rather than face the charges she fled to England to avoid jail leaving her own two small children behind. The charges were dropped upon her return to New York in 1916. Sanger continued to challenge the Comstock Laws and together with her sister and friends in 1916 founded the first birth control clinic in Brooklyn. In 1921, Sanger established the American Birth Control League, the antecedent of the Planned Parenthood Federation of America.

In 1917, Sanger met Katherine McCormick (1875–1967) in Boston at one of Sanger's lectures and the two struck up an enduring friendship. McCormick was an avid crusader for women's rights, was a leader in the Suffragette Movement, helped establish the League of Women Voters, and was the second woman to graduate from MIT with a degree in biology. With the death of her husband Stanley in 1947, McCormick became the heir to the fortune of International Harvester. In 1950, McCormick wrote to Sanger to ask how her inheritance might be used for contraceptive research and Sanger, in turn, frustrated with the slowing progress of the birth control movement decided on a new tactic—an oral contraceptive. McCormick first pledged $10,000 toward the research and thereafter began to contribute $150,000 to $180,000 a year through Planned Parenthood's research program. Ultimately, McCormick's gifts would total $2 million.

Sanger approached the biologist Gregory Pincus (1903–1967), who was a brilliant outcast from the scientific establishment. Pincus, a bushy-haired man with a gray mustache and dark, burning eyes, was a first-generation American born to Jewish immigrants Elizabeth Lipman and Joseph Pincus on April 9, 1903, in New Jersey. He went to Cornell University and received a bachelor's degree in agriculture in 1924. Then he went to Harvard University where he was an instructor in zoology while also working toward his master's and doctorate degrees. From 1927 to 1930, he moved from Harvard to Cambridge University in England to the Kaiser

Wilhelm Institute in Berlin, conducting research. He became an instructor in general physiology at Harvard in 1930, and was promoted a year later to assistant professor. By 1938, at the age of 35, he was already an international authority on the sex of mammals and sex hormones, and more than 70 of his research papers had been published. Though hired as a teaching professor at Harvard University, pure research and laboratory work was always a stronger draw. Early experiments in transferring eggs from one animal to another prefigured the surrogate mother model. Pincus's breakthrough research on the breeding of rabbits without males via artificial insemination in 1939 caused wide excitement, even outside scientific circles. However, other researchers were unable to reproduce the results. These irreproducible experiments with rabbits had a negative effect on Pincus attaining tenure at Harvard. Fired from Harvard, he began conducting research on rats and rabbits in a converted garage in Worcester, Massachusetts, using money raised by going door to door and asking for donations from housewives, plumbers, and hardware stores for a new institution he called the Worcester Foundation for Experimental Biology.

Sanger wanted Pincus to develop a foolproof method of contraception, preferably a pill, and he agreed to try. Pincus was an expert in mammalian reproduction and knew that progestins stopped ovulation in rabbits and rats and was eager to take the next step: women. For this, Pincus recruited the eminent gynecologist John Rock (1890–1984). Rock looked like a family physician from central casting in Hollywood. He was 6 ft 3 in. tall, rail thin, with a gentle smile and a calm and deliberate manner. With a shock of white hair, he stood straight as an arrow and had been teaching obstetrics at the Harvard Medical School for decades. He pioneered in vitro fertilization and freezing sperm. Furthermore, he was a Catholic who had dared to defy his Church by talking to young couples about sex and babies before marriage; he wanted them to understand that sex was neither shameful nor obscene, and he wanted married couples to have effective means for birth control—all tenets against the Church doctrine.

Beginning in the 1950s, Rock was treating women for "unexplained infertility," and he believed that the women were not conceiving because their reproductive systems were not fully developed. He hypothesized that when such infertile women did become pregnant the pregnancy helped the reproductive system to mature. To test the hypothesis, he recruited 80 frustrated but valiantly adventuresome women for an experiment in which he would use progesterone and estrogen, the same hormones Pincus had been studying. He started the women on 50 mg progesterone and 5 mg estrogen and found no harmful effects. Within a few months, 13 of the 80 women became pregnant when they stopped taking the hormones. During Rock's treatment, the women were convinced they were pregnant because the hormones produced the same effects as pregnancy: nausea, swollen breasts, and cessation of menstruation. Because preventing pregnancy was not the purpose of his treatment, he was able to skirt Massachusetts's law against contraception. However, when Rock told Pincus his patients were crushed by the mirage of pregnancy, Pincus suggested that instead of administering hormone therapy daily, it be administered for 21 days with a 7-day reprieve, allowing the subjects to experience regular menstruation. The benefit of this would be twofold: patients would have regular periods and be disabused of the notion that they were pregnant. Then Pincus presented Rock with another proposal: Would Rock allow some of his patients to take Pincus's oral birth control pill? The study would involve monitoring ovulation during their pseudo-pregnancies, and if they still benefitted from Rock's maturation effect, that would be great too. Rock agreed.

In August 1953, a year after Syntex had published on norethindrone, G. D. Searle filed a patent for the synthesis of a related compound, norethynodrel. Because Pincus was a consultant to Searle, he naturally picked the Searle compound for clinical evaluation. The first oral contraceptive study by John Rock, Gregory Pincus, and Pincus's Worcester Foundation collaborator Min-Chueh Chang involved 60 women, some of whom were infertility patients, whereas others were nurses. These small trials involved daily temperature readings, vaginal smears, and urine samples, as well as monthly endometrial biopsies. Although the initial results seemed promising in blocking ovulation, the sample size was small and few of the subjects complied with the protocol. More test subjects were needed.

Conducting more extensive clinical trials in the United States posed a challenge due to the Comstock Laws, so the first large-scale trials were conducted in Puerto Rico in 1956. Why Puerto Rico? Puerto Rico was densely populated; there were no laws against contraception; it had well-established birth control clinics; it was close enough to the United States for visits from the research team; the population was stable and could be followed for the full length of the trial; the physicians had been trained in the United States, and Pincus knew and trusted them; and there was a high demand for alternatives to permanent sterilization, which was widespread on the island. Test groups were composed of women who were for the most part less than 40 years of age, and they had to have had two children to prove they were fertile and agree to have the child if they became pregnant. After a propaganda campaign by a Catholic group, 10% of subjects dropped out of the study, and all of them became pregnant soon after. Test subjects' health was meticulously monitored. Side effects included nausea, headaches, and dizziness, reported by 17% of volunteers. These side effects were unacceptable, even though the pill proved completely effective in preventing pregnancy. In 1956, data on 221 subjects had been collected. The report stated, the drug "gives one hundred percent protection against pregnancy in 10-milligram doses taken for twenty days each month. … However, it causes too many side reactions to be acceptable generally." Rock wondered if the side effects were actually a placebo effect. A placebo-controlled trial seemed to confirm this, as 17% of subjects in the placebo group reported side effects, compared to the 6% of subjects who experienced side effects in an experimental group to whom no warning about side effects was given.

Critics of the Puerto Rico clinical trials point out retrospectively that the women, many of whom were illiterate or semi-illiterate, did not give informed consent with their signatures. But, it has to be noted that in the 1950s and early 1960s, it was not a common practice to have subjects sign informed consent documents to participate in a clinical trial. Today, it is a legal requirement of any research project involving human volunteers to have informed consent. It is also true that the number of women involved in the Puerto Rico clinical trial and amount of time they were observed would not pass muster today.

Pincus, a remarkable entrepreneur, persuaded Searle to market the drug, named Enovid, not for birth control but to help with "disturbances of menstruation." The U.S. Food and Drug Administration mandated a warning on the label that it would prevent ovulation. And women quickly learned that this pill would prevent pregnancy. By 1959, more than 500,000 American women had used Enovid for contraception. In 1960, the FDA approved the pill's use for birth control. (Enovid was discontinued in the United States in 1988 along with other first-generation high-estrogen combined oral contraceptive pills.)

In turn, Syntex, which had been excluded from clinical trials by Pincus and G. D. Searle, signed an exclusive license with Parke-Davis to market norethindrone. In 1957—the same year that Searle received FDA approval for Enovid for use in menstrual disorders and conditions of infertility— both drugs went on the market. Now a potential legal conflict between Searle and Syntex arose because the Searle compound (which was not covered by the Syntex patent) was transformed in the body to the Syntex compound; however, Syntex's partner Parke-Davis did not concur. (This was largely due to the fact that Searle was marketing the best-selling antinausea drug Dramamine that contained the Parke-Davis antihistamine Benadryl, and in 1957, norethindrone was not a block-buster drug; consequently Parke-Davis elected not to fight with a valued customer.)

Syntex's entrepreneurial counterpart to Pincus was Alejandro Zaffaroni, a Uruguayan with a PhD in biochemistry from the University of Rochester, who spearheaded the clinical trials for FDA approval of norethindrone as an oral contraceptive. Zaffaroni negotiated a licensing agreement with the Ortho division of Johnson & Johnson, but because Parke-Davis refused to release the results to Ortho of its monkey studies, Ortho had to repeat the work and so there was a delay in FDA approval to market Syntex's norethindrone (as Ortho-Novum) until 1962—2 years after Searle's contraceptive drug Enovid hit the market. By 1964, Parke-Davis had a change of heart and licensed the Syntex drug, and under the supervision of Zafferoni gained a major share of the U.S. oral contraceptive market. In 1994, Roche acquired Syntex. There was an ironic twist to the earlier competition between Syntex and Searle when Roche sold its entire line of oral contraceptives to Searle—the company that had gone to heroic efforts to circumvent the Syntex patent for norethindrone.

Aftermath

In 1952, Djerassi whose goal in life was to establish a scientific reputation, moved to Wayne State University in Detroit as an assistant professor, where he continued to work on steroids, the alkaloids of the giant cactus, and spectropolarimetry. In 1959, he joined Stanford University when a professor from graduate school days at Wisconsin, William S. Johnson, became the chair of its Chemistry Department. Djerassi had achieved his lifelong goal of moving from industry to academe. During his long career at Stanford University, he published more than 1000 research papers, taught a course "Biosocial Aspects of Birth Control," and formed a company Zoecon for nontoxic control of mosquitoes and fleas based on his synthesis of a synthetic juvenile hormone that prevented insects from becoming adults. The pill made him wealthy, but it was not because he received a royalty for each pill used by millions of women. For his discovery, he received a token payment of $1, but during his time at Syntex he also acquired stock in the company when they cost only $2 each; it was the stock flotation of Syntex in the 1960s on the U.S. Stock Exchange that made him a rich man and he bought 1200 acres in the Santa Cruz mountains near Stanford University where he created a colony for struggling artists and writers as well as becoming a collector of great art, particularly Paul Klee. In the final 25 years of his life, he wrote novels and plays in which scientists debated the ethical and social implications of what they were doing in the lab. He died peacefully surrounded by family and loved ones in his home in San Francisco at the age of 91.

In 1965, one out of every four married women in the United States used the pill and by 1967 13 million women in the world were using it. By 1984 that number reached 50–80 million, and today more than 100 million women use the pill. The first oral contraceptive, Enovid, had a lot more hormone than needed to prevent pregnancy: 10,000 µg of progestin (norethynodrel) and 150 µg of estrogen (mestranol) in comparison to today's pill that contains 50–150 µg of progestin and 20–50 µg of estrogen. The side effects noted in the first trials in Puerto Rico have been reduced but have not entirely eliminated occurrences of thromboembolism, stroke, heart attack, increased blood pressure, liver tumors, gallstones, headache, irregular bleeding, and nausea. Faced with these side effects women have to consider a risk benefit perspective in the use of the pill: the risk of thromboembolism in pregnancy is 30 in 10,000 and with the pill 9.1 in 10,000. The risk of mortality is the same in women taking the pill or not. Today, in a developed country a woman's chances of dying during pregnancy are 1 in 3000, and risk of complications from the pill are as low or lower than from other drugs and is equivalent to the risk of death a woman runs in her own kitchen in the same period of time. Currently, the pill is the most common method of birth control in the United States. Forty-four percent of U.S. women rely on the method for reasons other than preventing pregnancy including menstrual regulation, acne, endometriosis, and cramps.

Since its introduction, the pill has engendered controversy over promiscuity and the morality of birth control. In altering sexual and reproductive practices, it has affected family economics and the working lives of millions of women worldwide. By prolonging the age at which a woman first marries, it has allowed them to invest in education as well to become more career oriented. After the pill was legalized, there was a sharp increase in college attendance and graduation rates for women. By reducing the number of unwanted pregnancies, it has halved the number of maternal deaths per 100,000 live births, and infant deaths also declined because unwanted births are usually associated with delayed access to prenatal care and increased child abuse and neglect. The oral contraceptive pill has been named one of the Seven Wonders of the Modern World because as one writer stated, "When the history of the twentieth century is written it may be seen as the first time when men and women were truly partners. Wonderful things can come in small packets."

Dr. Frederick Banting (right) and Charles H. Best (left) and one of their dogs on the roof of the Medical Building, summer 1921. (From Banting's Scrapbook. With permission.)

Chapter
5

Diabetes and Insulin

Diabetes is a disease syndrome with at least 50 possible causes. The total number of diabetics in the United States is estimated to be 29 million or almost 10% of the population. There are more than 422 million people worldwide living with diabetes, and it contributes to 3.7 million deaths. Diabetes is the seventh leading cause of death in the United States, with a quarter of a million people dying from it as an underlying or contributing cause.

Disease Characteristics

Diabetes has been recognized as a disease for millennia. The first description of the symptoms of diabetes was in ~100 AD by the Greek physician Aretaeus of Cappadocia: "Diabetes ... being a melting down of the flesh and limbs into urine ... the patients never stop making water, but the flow is incessant, as if from the opening of aqueducts ... the disease is chronic and it takes a long period to form; but the patient is short lived ... life is ... painful; thirst unquenchable; and one cannot stop them from drinking or making water; nor does any great portion of the drink get into the system, and many parts of the flesh pass out along with the urine." The Greek physician Galen of Pergamon (129–200 AD) put it succinctly: it is a diarrhea of the urine. Another physician called it "the pissing evil." Although a standard part of ancient medical diagnosis was to taste the urine, it was not until the seventeenth century that Thomas Willis (1621–1675) remarked on the sweetness of the urine from a diabetic. A Liverpool physician, Matthew Dobson, who noted in 1772, that his patient was weak and emaciated and with an unquenchable thirst, established that the sweetness was due to sugar. When Dobson evaporated 2 quarts of the colorless urine it left a white residue whose taste was indistinguishable from sugar. In 1815 the French chemist Eugène Chevreul (1786–1889) used the newly developed chemical method for measuring the amount of sugar to show that the sugar in urine was glucose. Being able to measure the amount of glucose in the urine led some physicians to put their patients on special diets; most patients, however, did not adhere to the diet and the few who did were not cured. Patients longed for a drug that would supersede dieting, and by 1894, a U.S. government publication listed 42 antidiabetic remedies; all were worthless. In the 1880s, a time when the germ theory of Robert Koch and Louis Pasteur was in full flower, it was speculated that diabetes might be an infectious disease. However, no causative microbe was found in the urine or blood of diabetics.

Between 1846 and 1848, physiologist Claude Bernard (1813–1878) found that sugar was present in the blood of animals even when they were starved. Further, the sugar concentration was higher in the hepatic vein that leads from the liver to the general circulation than in the portal vein that takes blood from the intestine to the liver. He hypothesized that the sugar absorbed from the digested food in the intestine was stored in the liver as glycogen and then during fasting the glycogen was broken down into glucose and released into the bloodstream. Bernard maintained that the kidney was the dam that held back sugar until it exceeded a particular level, and when this higher level was reached, sugar would spill over into the urine. He went on to suggest that it was the liver that was involved in diabetes and when there was overproduction of sugar by the liver the person would show diabetic symptoms.

The Pancreas, an Endocrine Organ

The role of the pancreas itself was unknown until 1848 when Claude Bernard established that the gland produced digestive enzymes; however, its role in diabetes was unclear because tying off of the pancreatic duct in animals did not produce the disease. Then in 1889, the Lithuanian physician Oskar Minkowski (1858–1931) working in a Strasbourg University department where diabetes was the main subject of study and where experimentation was encouraged made a critical discovery: removal of the pancreas in dogs caused severe diabetes. Minkowski admitted that his discovery was a lucky accident. There was a spare dog available and he decided to remove its pancreas. Some days later the animal caretaker told him that the previously house-trained dog on which Minkowski had operated was urinating everywhere. Minkowski criticized the animal caretaker for not letting the dog out often enough to which the caretaker replied that letting it out made no difference. This prompted Minkowski to test the dog's urine. It was loaded with sugar. Minkowski went a step further. He cut the pancreas of a dog in half and transplanted one half into the abdominal wall where it took root. When he removed the remainder of the pancreas, diabetes did not develop. However, when the transplant was removed the dog became diabetic. He concluded that the pancreas must have been secreting something that prevented diabetes.

Bernard's finding that when there was overproduction of sugar by the liver the person would show diabetic symptoms was used as the basis for the treatment of diabetes by diet. The most successful of these diets was that developed by Frederick Allen (1876–1964) a square-faced, stern-looking man who never smiled. Between 1909 and 1912, Allen carried out experimental research on diabetes at Harvard University. Using dogs, he removed varying amounts of pancreas to produce the equivalent of severe or mild diabetes—what we would call today type 1 and type 2. Dogs left with 20% or more of their pancreas did not develop diabetes; however, the fate of those with 80%–90% of the pancreas removed depended on what the dogs ate. If fed a low carbohydrate diet the dogs remained relatively well, like middle-aged humans with type 2 diabetes. Extrapolating these results with dogs to humans Allen decreed that patients should order their lives according to the size of their pancreas, that is, reduce food intake until the urine is sugar free. In 1914, Allen was appointed to the staff of the Rockefeller Institute where he treated "hopeless" diabetics. In 1915, he published the results with 44 patients in a paper entitled "Prolonged Fasting in Diabetics" and the tabloid the *Daily Mail* heralded the findings: "Cure for Diabetes." Editorials in the *Lancet* and the *British Medical Journal*, however, were less enthusiastic. Allen's starvation diet worked to a limited degree and was welcomed by some physicians because that was all they had to offer. Although Allen and other supporters of his diet claimed their failures were due to "unfaithfulness on the part of the patient," the truth is that many of the patients died from starvation not diabetes.

In 1894, Gustave-Édouard Laguesse (1861–1927) following Minkowski's work on the pancreas suggested that secretions came from the patches of small irregular polygonal-shaped cells that stained a bright red (with aniline dyes) and were scattered throughout the pancreas like islands in a sea; these had been discovered in Berlin in 1869 by a medical student, Paul

Langerhans (1847–1889), and are still called the islets or islands of Langerhans. In 1899, when Leonid Sobolov (1876–1919) found that in 4 out of 15 cases of diabetes the islets had disappeared he suggested that the islets were independent of the rest of the pancreas and controlled the metabolism of carbohydrates. And like Sobolov in 1901, Eugene Opie (1873–1971) at Johns Hopkins thought the islets were organs of internal secretion, that is, ductless endocrine glands. He suggested that damage to the islets would result in diabetes and the finding of obliterated islets in autopsies of juvenile diabetics supported this. Now the challenge was to isolate and characterize the internal (endocrine) secretions (=hormones) produced by the islets of Langerhans.

Discovery of Insulin

In 1923, J. J. R. Macleod and F. G. Banting were awarded the Nobel Prize for Physiology or Medicine for "the discovery of insulin" and since that time it has been a source of controversy. It is not for the importance of the discovery itself that controversy has raged rather the dispute has been why Macleod and Banting were singled out for the honor.

Banting, who received a medical degree from the University of Toronto, was wounded as a medical officer in World War I, and by 1919 was a resident in surgery at the Hospital for Sick Children in Toronto, Canada. After missing out on a prestigious appointment at the hospital he took a part-time job at the University of Western Ontario where he had to lecture students on carbohydrate metabolism of which he knew little. While preparing his lectures he read about a case in which a stone blocked the pancreatic duct leading to atrophy of the cells of the pancreas that produce the digestive enzymes, but left the islets of Langerhans intact. This was not a new finding since it was well known to physiologists that this is what happened when the duct was ligated in experimental animals. Banting later recalled that during the early morning hours of October 31, 1920, he wrote in his notebook: "Ligate pancreatic duct of dog. Keep alive till acini degenerate leaving islets. Try to isolate the internal secretion of these to relieve glycosuria (sugar in urine)." To pursue this idea experimentally he consulted with J. R. Macleod, Professor of Physiology at the University of Toronto on November 8, 1920. At first, Macleod was unimpressed by both Banting and the idea, but when Banting approached Macleod again in March of the following year, writing that he would like to spend half of May plus the summer in Macleod's laboratory, he agreed. Together they worked out a plan of investigation with Macleod providing experimental animals (dogs) but no salary. Macleod suggested removal of the pancreas from dogs to produce diabetes, demonstrated how to ligate the duct, and then to assist in the measurement of glucose in the urine and blood provided him with a medical student, Charles Best. Macleod allowed Banting to use a dirty little room in the physiology department that had been used for surgical research. In June, before Macleod left for a summer vacation in Scotland, he gave explicit instructions on how the experiments should proceed. By the end of July, Banting and Best had experimented on 19 dogs; 14 had died and 5 had tied ducts but only 2 were suitable for experimentation. Banting removed the atrophied pancreas from one ligated dog, made a saline extract from the shriveled pancreas, and injected some of it intravenously; the blood sugar level dropped. The procedure was repeated on a second dog with similar results. Believing they had identified the secretion they named the extract "Isletin." The work continued into August with similar findings. When Macleod returned in September he wanted more work to be done; Banting in turn asked for a salary, a better room, and more dogs; he threatened to leave if these were not provided. Macleod was incensed by the confrontation. But later he relented and by October, Banting and Best were on the University payroll. In November, Macleod asked Banting and Best to talk to the students and faculty about their work at the meeting of the Physiological Journal Club. Banting had asked Macleod to introduce them and when he did so he said all the things that Banting had intended to say about their summer research. Afterward, Banting became angry especially when Macleod used the pronoun "we" in describing the work. He was further irritated when he discovered that the students in attendance had talked about the remarkable work of Professor Macleod. This continued to foster Banting's antipathy toward Macleod.

With only a few ligated dogs left to work with, Banting and Best opted to make the extract from a calf pancreas obtained from the local abattoir. (The idea of making extracts had come from Macleod not Banting.) They improved on the extraction procedure, and in November, their results were reported in a paper "Internal Secretion of the Pancreas" in the *Journal of Laboratory and Clinical Medicine*. In the publication, they omitted mention of the disappointing summer experiments and they also exaggerated the results stating that "the extracts always produced a reduction in sugar" in the diabetic dogs. Indeed, of 75 injections of supposedly degenerated or "exhausted" pancreas using 9 dogs there were 42 favorable, 22 unfavorable, and 11 inconclusive results. The next breakthrough came when Banting and Best used alcohol in the preparation of the extract. They used fetal calf pancreas presumably because it was richer in islets cells than adult pancreas. It worked well. By December, they had abandoned using ligation and instead of throwing out the dog's pancreas after pancreatectomy they used fresh adult dog pancreas to make extracts. It also worked well in reducing blood sugar. Running low on normal dog pancreas they were led to use adult beef pancreas as a more abundant source for making extracts. It must have been a surprise that it too worked so well since the original rationale for ligation of the dog pancreatic duct was that the internal secretion would be destroyed by the protein-digesting enzymes from the pancreas. In fact, it is clear Banting and Best were unaware that as early as 1875, a German physiologist had shown that fresh pancreas did not have any protein-digesting enzyme activity. (The intact pancreas contains an inactive form of the protein-digesting enzyme, trypsinogen, which is converted into the active form, trypsin, only after contact with the digestive juices in the small intestine.)

In December 1921 at the American Physiological Society meeting held at Yale University, Macleod, Banting, and Best presented a paper "The Beneficial Influences of Certain Pancreatic Extracts on Pancreatic Diabetes." Banting was nervous and spoke haltingly; he did not convince the audience that the results proved the presence of an internal secretion by the pancreas. Was it the extract or was it the remnants of the pancreas that kept the dogs alive? Were routine autopsies adequate to prove total pancreatectomy? Instead of Banting responding to these questions Macleod came to his defense. Banting was rankled that Macleod took over and defended the work so smoothly. Banting decided that the professor had gone too far; Macleod kept using the word "we" even though he had not done a single experiment. Banting felt Macleod was trying to steal his results. Macleod however knew nothing of Banting's feelings.

Back in Toronto Banting pressed Macleod for help in purifying the pancreas extract and in 1921 James B. Collip a PhD in biochemistry from the University of Toronto joined the effort. Collip, made extracts from whole beef pancreas using increasing concentrations of alcohol and found the active principle remained in solution at higher concentrations when most of the proteins precipitated out. Then, at somewhere over 90% alcohol the active principle precipitated out. The material still contained impurities but it was the most active and purer than any other preparation. In January 1922, Collip found that pancreatic extracts also lowered the blood sugar of normal rabbits. The rabbit assay was a quick and easy way of determining the potency of a batch of extract. On occasion, Collip's extract produced hunger in the rabbits and as their blood sugar dropped they started to eat paper or wood shavings; with the most potent batches the rabbits went into convulsions and collapsed into coma. With this, Collip had discovered what we now call "insulin shock." Faced with these findings, Macleod turned his whole laboratory over to the problem with Collip, not Banting and Best, preparing extracts. Collip became less and less communicative with Best and Banting, and Collip and Macleod decided not to tell Banting and Best the secret of making an effective antidiabetic extract.

On January 23, 1922, Collip's extract was administered to Leonard Thompson a severely diabetic 14 year old; glycosuria disappeared and his blood sugar became normal. By the end of February there was enough clinical and experimental evidence to support a preliminary publication entitled "Pancreatic Extracts in the Treatment of Diabetes Mellitus." Banting, listed as an author, had little to do with the writing of the paper or the clinical work. Years later, an embittered Banting wrote "Best and I became technicians under Macleod like the others. Neither plans for experiments nor results were discussed with us." By April 1922, the Toronto group prepared a paper, "The Effect Produced on Diabetes by Extracts of Pancreas," summarizing the work thus far. It included Banting's idea, Banting and Best's early experiments, and Collip's purification and clinical results. For the first time a name was given to "this extract which we propose to call insulin." On May 3, 1922, at a meeting of the Association of American Physicians, Macleod speaking for the Toronto group announced to the medical world that they had discovered the hormone insulin. It was a great triumph that had begun on May 17, 1921, by Banting and Best.

Insulin, the Drug

On April 12, 1922, a patent on the insulin process was taken out in the names of Collip and Best and immediately assigned to the Board of Governors of the University of Toronto to stop anyone from ever being in a position to stop anyone else from making the extract. In North America the University of Toronto gave Eli Lilly and Company exclusive rights to produce and sell insulin (first as Isletin and later as Insulin Lilly) at cost, but even at this price it was expensive. By mid-September 1923, 20,000–25,000 diabetics in the United States were receiving

Lilly insulin from 7,000 physicians. In its first year of marketing, Lilly sold a million dollars worth of insulin. More than any other product insulin transformed the company into a giant in the American pharmaceutical industry.

In October 1922, the Danish Nobel laureate August Krogh asked Macleod if he could test insulin in his country since his wife had developed diabetes a year earlier. By 1923, using vast supplies of pork pancreas available from Denmark's bacon factories the Nordisk Insulin Laboratory was manufacturing Insulin-Leo. Connaught Laboratories solely owned by the University of Toronto supplied Canada's insulin. In 1972, the University sold Connaught and put the proceeds in a fund to support research. Connaught's insulin division was sold to Novo-Nordisk the company that evolved from the Nordisk Insulin Laboratory.

The need for a long-acting insulin became obvious early on when purification of ordinary or soluble insulin shortened its action and meant that people needed an injection before each meal and again at 3 a.m. For prolonged action, protamine insulinate was made at the Nordisk Insulin Laboratory in 1936. Later, protamine zinc insulin was developed in Toronto and its selling point was that it lasted 24 h; it was prescribed for type 2 diabetes, but for type 1 diabetes soluble insulin had to be added. The most popular insulin developed in 1940 by Nordisk was neutral protamine Hagedorn or isophane. The Lente series, a particulate combination of zinc and insulin was developed in 1952 that required only a once a day injection. In the 1980s, Eli Lilly and then Novo-Nordisk started selling genetically engineered human insulin and soon it was possible to make "designer" insulin and insulin analogs that would facilitate insulin's action and make control of diabetes easier.

What Happened to Those Who Discovered Insulin?

Banting was so incensed at having to share the 1923 Nobel Prize for Physiology or Medicine with Macleod that at first he was tempted to refuse it, and then when he did accept it, he shared the prize money with Best. Macleod in turn shared his prize money with Collip. Macleod left the University of Toronto in 1928 and became Regius Professor at the University of Aberdeen (his alma mater) in Scotland. When questioned about the discovery of insulin he politely turned aside each and every inquiry. He died in 1935 at age 59. Banting continued to believe in his own myth and felt he deserved all the recognition he got. He continued in his hatred of Macleod and was paranoid in believing he had been deprived of full credit for his work on insulin. He became a well-to-do research professor with no teaching obligations at the University of Toronto. After insulin most of his research was directed at finding a cure for cancer. His ideas did not pan out. In 1934, he was honored with a knighthood. He died in a plane crash in 1941. Best completed his medical education and after Banting died became the head of the Banting and Best Department at the University of Toronto. In later years, he was showered with honors. He retired in 1967, and died in 1978. Neither Banting nor Best gave much thought to Collip. Collip in turn went out of his way in his academic publications to identify himself as a co-discoverer of insulin. Collip eventually earned a degree in medicine, went on to isolate parathyroid hormone, became chair of the Department of Biochemistry at McGill University, and in 1947 was appointed the Dean of Medicine at the University of Western Ontario. He died in 1965 at age 72.

The story of the discovery of insulin makes clear that Banting and Best alone did not discover insulin. The identification of the internal secretion of the pancreas—the hormone insulin—was a multistep process involving the vital contributions of many. Indeed, it is important to note that Banting's idea of duct ligation played no essential role in the discovery. What Banting did provide was determination and an unshakable albeit unscientific faith in the quest. Banting and Best were not experienced enough in physiology or biochemistry to have carried the work through to a successful conclusion and for that they badly needed the counsel of Macleod. Furthermore, it is unlikely that were it not for Macleod's suggestions, Banting and Best would not have made pancreas extracts or used alcohol for purification. Macleod's failure was that as a mentor he was not critical of the flaws in the experiments of Banting and Best. But to his credit, he finally did see the importance of the work and was then able to add his and Collip's scientific acumen as well the efforts of clinicians W. R. Campbell and A. A. Fletcher to the team. Robert Tattersall has written "The tragedy of the interaction was that the compound of powerful personalities necessary to produce the great scientific advance was so unstable. The team was impossibly volatile … It was … a problem in human relations."

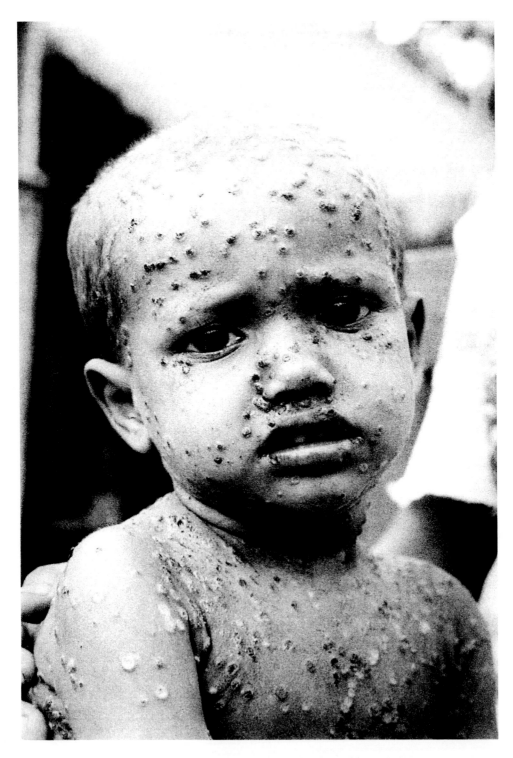

A child with smallpox. (Courtesy of the CDC Public Health Image Library.)

Chapter 6

Smallpox and Vaccination

Smallpox is no longer with us, but over the centuries, it killed hundreds of million people. In the twentieth century alone, it killed 300 million people—three times the number of deaths from all the twentieth-century wars. Smallpox has been involved with war, with exploration, and with migration. As a result, it changed the course of human history. Smallpox was indiscriminate, with no respect for social class, occupation, or age; it killed or disfigured princes and paupers, kings and queens, children and adults, farmers and city dwellers, generals and their enemies, and rich and poor.

A Look Back

At one time, smallpox was one of the most devastating of all human diseases, yet no one really knows when smallpox began to infect humans. It is suspected that humans acquired the infectious agent from one of the pox-like diseases of domesticated animals, in the earliest concentrated agricultural settlements of Asia or Africa, when humans began to maintain herds of livestock, sometime after 10,000 BC. The best evidence of smallpox in humans is found in three Egyptian mummies, dating from 1570 to 1085 BC, one of which is Pharaoh Ramses V, who died as a young man in 1155 BC. The mummified face, neck, and shoulders of the Pharaoh bear the telltale scars of the disease: pockmarks. (The word pox comes from the Anglo-Saxon word "pocca," meaning pocket or pouch and describes the smallpox fingerprint—craterlike scars.)

From its origins in the dense agricultural valleys of the great rivers in Africa and India, smallpox spread from the West to China, first appearing about 200 BC. Trade caravans assisted in the spread of smallpox, but at the time of the birth of Christ, it was probably not established in Europe because the populations were too small and greatly dispersed. Smallpox was unknown in Greece and Rome and was not a major health threat until about 100 AD, when there was a catastrophic epidemic called the Plague of Antoninus. The epidemic started in Mesopotamia, and the returning soldiers brought it home to Italy. It raged for 15 years, and there were 2000 deaths daily in Rome.

There are records of smallpox in the Korean peninsula from 583 AD, and 2 years later, it reached Japan. In the western part of Eurasia, the major spread of smallpox occurred in the eighth and ninth centuries during the Islamic expansion across North Africa and into Spain and Portugal. As the disease moved into Central Asia, the Huns were infected either in Persia or in India. In the fifth century, when the Huns descended into Europe, they might have carried smallpox with them. By 1000 AD, smallpox was probably endemic in the more densely populated parts of Eurasia, from Spain to Japan, as well as in the African countries bordering on the Mediterranean Sea. The movement of people to and from Asia Minor during the Crusades in the twelfth and thirteenth centuries helped reintroduce smallpox to Europe. By the fifteenth century, smallpox was established in Scandinavia, and by the sixteenth century, it was established in all of Europe, except Russia.

Smallpox was a serious disease in England and Europe during the sixteenth century. As British and European explorers and colonists moved into the newly discovered continents of the Americas, Australia, and Africa, smallpox went with them. Smallpox played a crucial role in the Spanish conquests in the New World, and it also contributed to the settlement of North America by the French and English. The English settlers who arrived in the American colonies in 1617 set off an epidemic among the Indians, thereby clearing a place for the settlers who came from Plymouth in 1620. The English used smallpox as a weapon of germ warfare. In the War of 1763 between England and France for control of North America, under orders of Sir Geoffrey Amherst (Commander-in-Chief of North America), the British troops were asked: "Could it not be contrived to send the smallpox among those disaffected tribes of Indians? We must on this occasion, use every stratagem in our power to reduce them." The ranking British officer for the Pennsylvania frontier, Colonel Henry Bouquet, wrote back: "I will try to inoculate the Indians with some blankets that may fall into their hands and take care not to get the disease myself." Blankets were deliberately contaminated with scabby material from the smallpox pustules and delivered to Indians, allowing the spread of the disease among these highly susceptible individuals. This led to extensive numbers of deaths and ensured their defeat.

In West Africa, smallpox traveled with the caravans that moved from North Africa to the Guinea Coast. In 1490, the Portuguese spread the disease into the more southerly regions of West Africa. Smallpox was first introduced in 1713 into South Africa by a ship that docked in Cape Town that carried contaminated bed linens from a ship returning from India. The first outbreak of smallpox in the Americas was among African slaves on the island of Hispaniola in 1518. Here and elsewhere, the Amerindian population was so decimated that by the time of Pizzaro's conquest of Peru, it is estimated that 200,000 people had died. With so much of the Amerindian labor force lost to disease, there was an increased need for replacements to do the backbreaking work in the mines and plantations of the West Indies, the Dominican Republic, and Cuba. This need for human beasts of burden stimulated, in part, the slave trade from West Africa to the Americas. Slaves also brought smallpox to the Portuguese colony of Brazil. Smallpox was first recorded in Sydney, Australia, in 1789, a year after the British had established a penal colony there. It soon decimated many of the aboriginal tribes. As in the Americas, this destruction of the native population by smallpox paved the way for easy expansion by the British.

The Disease Smallpox

The cause of smallpox is a virus, one of the largest of the viruses, and with proper illumination, it can actually be seen with the light microscope. However, much of its detailed structure can be visualized only by using the electron microscope. The outer surface (capsid) of the smallpox virus resembles the facets of a diamond, and its inner dumbbell-shaped core contains the genetic material, double-stranded DNA. The virus has about 200 genes, 35 of which are believed to be involved in virulence.

Most commonly, the smallpox virus enters the body through droplet infection by inhalation. However, it can also be acquired by direct contact or through contaminated fomites (inanimate objects) such as clothing, bedding, blankets, and dust. The infectious material from the pustules can remain infectious for months. In the mucous membranes of the mouth and nose, the virus multiplies. During the first week of infection, there is no sign of illness; however, the virus can be spread by coughing or by nasal mucus. The virus moves on to the lymph nodes and then to the internal organs via the bloodstream. Here, the virus multiplies again. Then, the virus reenters the bloodstream. Around the ninth day, the first symptoms appear: headache, fever, chills, nausea, muscle ache, and, sometimes, convulsions. The person feels quite ill. A few days later, a characteristic rash appears.

In *The Demon in the Freezer*, Richard Preston described it:

> The red areas spread into blotches across (the) face and arms, and within hours the blotches broke out into seas of tiny pimples. They were sharp feeling, not itchy, and by nightfall they covered (the) face, arms, hands, and feet. Pimples were rising out of the soles of (the) feet and on the palms of the hands, too. During the night, the pimples developed tiny, blistered heads, and the heads continued to grow larger ... rising all over the body, at the same speed, like a field of barley sprouting after the rain. They hurt dreadfully, and they were enlarging into boils. They had a waxy, hard look and they seemed unripe ... fever soared abruptly and began to rage. The rubbing of pajamas on (the) skin felt like a roasting fire. By dawn, the body had become a mass of knob-like blisters. They were everywhere, all over ... but clustered most thickly on the face and extremities. ... The inside of the mouth and ear canals and sinuses had pustulated ... it felt as if the skin was pulling off the body, that it would split and rupture. The blisters were hard and dry, and they didn't leak. They were like ball bearings embedded in the skin, with a soft velvety feel on the surface. Each pustule had a dimple in the center. They were pressurized with an opalescent pus.
>
> The pustules began to touch one another, and finally they merged into confluent sheets that covered the body, like a cobblestone street. The skin was torn away ... across ... the body, and the pustules on the face combined into a bubbled mass filled with fluid until the skin of the face essentially detached from its underlayers and became a bag surrounding the tissues of the head ... tongue, gums, and hard palate were studded with pustules ... the mouth dry. ... The virus had stripped the skin off the body, both inside and out, and the pain ... seemed almost beyond the capacity of human nature to endure.

The individual is infectious a day before the rash appears and until all the scabs have fallen off. Many die a few days or a week after the rash appears. Not infrequently, there are complications from secondary infections. The infection results in a destruction of the sebaceous (oil) glands of the skin, leaving permanent craterlike scars on the skin—the telltale pockmarks.

By the beginning of the eighteenth century, nearly 10% of the world's population had been killed, crippled, or disfigured by smallpox. The eighteenth century in Europe has been called the "age of powder and patches," because pockmarks were so common. The "beauty patch" (a bit of colored material) seen in so many portraits was designed to hide skin scars. In this way, the pockmark set a fashion.

"Catching" Smallpox

Smallpox is a contagious disease and spreads from person to person. There are no animal reservoirs; that is, it is not a zoonotic disease. It can exist in a community only as long as there are susceptible humans. Smallpox caused large epidemics when it first arrived in a virgin community, where everyone was susceptible; afterward however, most individuals who were susceptible, either recovered and were immune or died; then, the disease died down. If the infection was reintroduced later, that is, after a new crop of susceptible individuals had been born or migrated into the area, then another epidemic would occur. However, after a period of time, the disease would reappear so frequently that only newborns would be susceptible to infection, older individuals being immune from previous exposure. In this situation, smallpox became a disease of childhood, similar to measles, chicken pox, mumps, whooping cough, and diphtheria.

Variolation

Early treatments of smallpox involved prayer and quack remedies. In ancient Africa and Asia, there were smallpox gods and goddesses that could be enlisted for protection. Because smallpox produces a red rash, a folklore remedy of "like cures like" arose in medieval times. In 1314, the Englishman John of Gaddeson followed this recipe and recommended that a smallpox victim could be helped by the color red, and so, those who were infected were dressed in red. Other charlatans suggested that cure would come about by eating red foods and imbibing red drinks. These remedies, which did no good, persisted for some quarters until the 1930s! Today, prevention of smallpox is often associated with Edward Jenner, but even before Jenner's scheme of vaccination, techniques were developed to induce a mild smallpox infection. The educated classes called this practice of folk medicine "inoculation," from the Latin word "inoculare," meaning "to graft," because it was induced by cutting. It was also called "variolation," since the scholarly name for smallpox was "variola" from the Latin word "varus," meaning "pimple." Possibly, the method of protection by variolation may have been inspired by the folklore belief of "like cures like," since in some versions of this remedy, secretions from an infected individual would be inoculated into the individual to be protected from that same infection.

There were numerous techniques of variolation. The Chinese avoided contact with the smallpox-infected individual, preferring that the person either inhaled a powder from the dried scabs shed by a recovering patient or placed the powdered scabs on a cotton swab and inserted this into the nostrils. Thomas Bartholin, physician to Charles V, King of Denmark and Norway, wrote the first scholarly account of variolation in 1675. Forty years later, this work came to the notice of the physician to Charles XII, Emanuel Timoni, who lived in Turkey. He, in turn, communicated a description of their method of variolation to the Royal Society of London: material was taken from a ripe pustule and rubbed into a scratch or into an incision in the arm or leg of the person being inoculated.

Mary Pierpont was a highborn English beauty who eloped with Edward Wortley Montagu, a grandson of the first Earl of Sandwich. After Edward became a member of Parliament, the couple soon became favorites at the court of King George I. In London in 1715, at the age of 26, Lady Mary Montagu contracted smallpox. She recovered, but smallpox-induced facial scars and the loss of her eyelashes marred her beauty forever. However, her 22-year-old brother was less fortunate and died from smallpox. The following year, Lady Montagu's husband was appointed ambassador to Turkey, and she accompanied him. There, she learned about the Turkish practice of variolation. She was so impressed by what she saw that she had her son inoculated. On returning to England in 1718, she propagandized the practice, and she had her daughter inoculated against smallpox at the age of 4. She convinced Caroline, the Princess of Wales (later Queen of England during the reign of George II), that her children should be variolated, but before that could happen, the royal family required a demonstration of its efficacy. In 1721, the "Royal Experiment" began with six prisoners who had been condemned to death by hanging and were promised their freedom if they subjected themselves to variolation. They were inoculated and suffered no ill effects, and on recovery, they were released, as promised. Despite this, Princess Caroline wanted more proof and insisted that all the orphan population of St. James Parish be inoculated. In the end, only six children were variolated, with great success. Princess Caroline, now convinced, had both her children (Amelia and Caroline) variolated. This too was successful. However, Lady Montagu met with sustained and not entirely inappropriate opposition to the practice of variolation. The clergy denounced it as an act against God's will, and physicians cautioned that the practice could be dangerous to those inoculated as well as others to those with whom they came in contact. The danger of contagion was clearly demonstrated when the physician who had previously inoculated Lady Montagu's children variolated a young girl, who then proceeded to infect six of the servants in the house; although all seven recovered, it did cause a local epidemic. Although inoculation was embraced by the upper classes in England and Lady Montagu's physician was called upon to variolate Prince Frederick of Hanover, it never caught on with the general population there or elsewhere. This had little to do with some of its demonstrated successes and more with the fear of contagion, the risks involved in inoculation with smallpox, and the lack of evidence that the protection would be long lasting. Later, it was found that variolation did provide lifelong protection, but the danger from death remained at ~2%.

Variolation came into practice quite independently in the English colonies in America. Cotton Mather, a Boston clergyman and scholar, was a member of the Royal Society of London. Although he read the account of Timoni, he wrote to the Society that he had already learned of the process of variolation from one of his African slaves, Onesimus. In April 1721, there was a smallpox outbreak in Boston. Half of the inhabitants fell ill, and mortality was at 15%. During this time, Mather tried to encourage the physicians in the city to carry out variolation of the population. The response from the physicians was negative, except for one, Zabdiel Boylston. On June 26, 1721, Boylston variolated his 6-year-old son, one of his slaves, and the slave's 3-year-old son, without complications. By 1722, Mather, in collaboration with Dr. Boylston, had inoculated 242 Bostonians and found that it protected them. His data showed a mortality rate of 2.5% in those variolated compared with the 15%–20% death rate during epidemics. However, Boylston's principal opponents were not the clergy but his fellow physicians.

In the crowded cities of England, during the eighteenth century, most adults had acquired some measure of immunity to smallpox through exposure during childhood. However, in the sparsely settled villages in the American colonies, there was little exposure and hence little immunity. With the outbreak of war in 1775, the most dangerous enemy that the colonists had to fight was not the British but smallpox. Following the Battle of Bunker Hill in June 1775, General Howe's forces occupied Boston, and Washington's troops were deployed on the surrounding hills. Howe was unable to attack, because smallpox was so rampant among the Bostonians and, to a lesser degree, among his troops (perhaps because of their practice of variolation). However, Washington did not attack for fear that his troops would be decimated by smallpox. Therefore, he first ordered his troops to be variolated. When the British evacuated the city on March 17, 1776, Washington sent 1000 of his men who were immune to smallpox to take possession of the city.

Earlier (in 1775), the American colonists feared that the British forces from Quebec, Canada, would attack New York. The Americans sent 2000 colonial troops under General Benedict Arnold to attack the poorly fortified garrison at Quebec. Although General Arnold's army was superior and better equipped, the colonials halted their advance on Quebec when they were suddenly stricken with smallpox. Over half of Arnold's troops developed the disease and mortality was very high. Few of the others were well enough to fight. In contrast, the much smaller British forces were well variolated and fit, so they could hold out until reinforcements arrived. The colonials abandoned their siege of the garrison and retreated to Ticonderoga and Crown Point on Lake Champlain, where they continued to die at a high rate. The Americans never made another serious attempt to intrude into Canada. Ten years later, based on this experience, Washington ordered that his entire army be variolated, but by that time, it was too late—Canada had been preserved for the British Empire.

Vaccination

The eradication of smallpox came about, thanks to the development of the first vaccine by an English physician, Edward Jenner (1749–1823). Jenner did not know what caused smallpox or how the immune system worked; nevertheless, he was able to devise a practical and effective method for immune protection against attack by the lethal virus *Variola major*. Essentially, Jenner took advantage of a local folktale and turned it into a reliable and practical device against the ravages of smallpox. The farmers of Gloucestershire believed that if a person contracted cowpox they were assured of immunity to smallpox. In milk cows, cowpox shows up as blisters on the udders, which clear up quickly and produce no serious illness. Farm workers who come in contact with cowpox-infected cows develop a mild reaction, with the eruption of a few blisters on the hands or lower arm. Once exposed to cowpox, neither cows nor humans develop any further symptoms. It was noted that a milkmaid or a farmer who had contracted cowpox hardly showed any of the disfiguring scars of smallpox and most milkmaids were reputed to have fair and almost perfect complexions, as suggested in the nursery rhyme:

> Where are you going, my pretty maid
> I'm going a-milking sir, she said
> May I go with you my pretty maid
> You're kindly welcome, sir, she said

What is your father, my pretty maid
My father's a farmer, sir, she said
What is your fortune, my pretty maid
My face is my fortune, sir, she said.

In 1774, a cattle breeder named Benjamin Jesty contracted cowpox from his herd, thereby immunizing himself. He then deliberately inoculated his wife and two children with cowpox. They remained immune even 15 years later when they were deliberately exposed to smallpox. However, Jenner, not Jesty, is given credit for vaccination, because he carried out experiments in a systematic manner over a period of 25 years to test the farmer's tale. Jenner wrote: "It is necessary to observe, that the utmost care was taken to ascertain, with most scrupulous precision, that no one whose case is here adduced had gone through smallpox previous to these attempts to produce the disease. Had these experiments been conducted in a large city, or in a populous neighborhood, some doubts might have been entertained; but here where the population is thin, and where such an event as a person's having had the smallpox is always faithfully recorded, no risk of inaccuracy in this particular case can arise."

As with the earlier "Royal Experiment," Jenner felt free to carry out human experimentation without any reservations, although it should be noted that all of his patients wanted the vaccination for themselves or for their children. On May 14, 1796, he took a small drop of fluid from a pustule on the wrist of Sarah Nelms, a milkmaid who had an active case of cowpox. Jenner smeared the material from the cowpox pustule onto the unbroken skin of a small boy, James Phipps. Six weeks later, Jenner tested his "vaccine" (from the Latin word "vacca," meaning "cow") and its ability to protect against smallpox by deliberately inoculating the boy with material from a smallpox pustule. The boy showed no reaction—he was immune to smallpox. In the years that followed, "poor Phipps" (as Jenner called him) was tested for immunity to smallpox about a dozen times, but he never contracted the disease. Jenner wrote up his findings and submitted them to the Royal Society of London, but to his great disappointment, his manuscript was rejected. It has been assumed that the reason for this was that he was simply a country doctor and not a part of the scientific establishment. In 1798, after a few more years of testing, he published a 70-page pamphlet "An Inquiry into the Causes and Effects of Variola Vaccinae," in which he reported that the inoculation with cowpox produced a mild form of smallpox that would protect against severe smallpox, as did variolation. He correctly observed that the disease produced by vaccination was so mild that the infected individual would not be a source of infection to others—a discovery of immense significance.

The reactions to Jenner's pamphlet were slow, and many physicians rejected his ideas. Cartoons appeared in the popular press, showing children being vaccinated and growing horns. However, within several years, some highly respected physicians began to use what they called the "Jennerian technique," with great success. By the turn of the century, the advantages of vaccination over variolation were clear, and Jenner became famous. In 1802, Parliament awarded him a prize of 10,000 pounds and another 20,000 pounds in 1807. Napoleon had a medal struck in his honor, and he received honors from governments around the world.

With general acceptance of Jenner's findings, the hide of the cow called "Blossom," which Jenner used as the source of cowpox in his experiments, was enclosed in a glass case and placed on the wall of the library in St. George's Hospital in London, where it remains to this day.

By the end of 1801, Jenner's methods came into worldwide use and it became more and more difficult to supply sufficient cowpox lymph or to ensure its potency when shipped over long distances. To meet this need, Great Britain instituted an Animal Vaccination Establishment, in which calves were deliberately infected with cowpox and lymph was collected. Initially, this lymph was of variable quality, but when it was found that addition of glycerin prolonged preservation, this became the standard method of production. The first "glycerinated calf's lymph" was sent out in 1895.

Why does vaccination work? There are two species of smallpox virus in humans: *V. major*, which can kill, and the less virulent, *V. minor*. Variolation or inoculation with *V. minor* obviously protects against the lethal kind of virus. However, vaccination involves cowpox, caused by a different virus named by Jenner as *Variolae vaccinae*. The exact relationship between the virus used in vaccines today and Jenner's cowpox is obscure. Indeed, in 1939, it was shown that the viruses present in the vaccines used then were distinct from contemporary cowpox as well as smallpox. Samples of Jenner's original vaccine are not available, so it cannot be determined what he actually used, but current thinking is that the available strains of vaccine were derived neither from cowpox nor from smallpox but from a virus that became extinct in its natural host: the horse.

Variolae vaccinae and *Variola major* are 95% identical and differ from each other by no more than a dozen of the 200 predicted genes in the genome. In order for there to be protection, there must be cross-reactivity between the two, such that antibodies are produced in the vaccinated human that can neutralize the *V. major* virus, should the need arise. Fortunately, this does happen.

Smallpox is a disease of historical interest, because it was certified as eradicated by the World Health Assembly on May 8, 1980. This feat was accomplished 184 years after Jenner had introduced vaccination. The smallpox eradication program began in 1967. By 1970, smallpox had been eliminated from 20 countries in western and central Africa; by 1971, it had been eliminated from Brazil; by 1975, it had been eliminated from all of Asia; and in 1976, it had been eliminated from Ethiopia. The last natural case was reported from Somalia in 1977. Eradication of smallpox was possible for three reasons: First, there were no animal reservoirs. Second, the methods of preserving the vaccine proved to be effective. Third, the vaccine was easily administered. Smallpox is the first and only naturally occurring disease to be eradicated by human intervention.

The Social Context of Smallpox Vaccination

In 1799, news of Jenner's vaccine reached the American colonies. Benjamin Waterhouse (1754–1846), a physician and professor at the newly established Harvard Medical School, was one of the first to send for the vaccine, and he popularized its use in Boston. His first trial was in 1800, when he vaccinated his son; then, he sent a pamphlet and a list of successes to Vice President Thomas Jefferson. Jefferson enthusiastically endorsed the "Jennerian technique" and had his own family and his slaves vaccinated. Jefferson expected the public to follow his lead but they did not. Indeed, it was not until the term of James Madison, the fourth president of the United States, who was familiar with the positions of Waterhouse and Jefferson, that a vaccine law was signed, encouraging vaccination. However, Waterhouse's success with Jefferson turned out to be grist for the mill of his adversaries. Waterhouse, a practicing Quaker, was a pacifist and opposed the Revolutionary War. Born in Rhode Island, he was considered an outsider to the conservative Boston community, and he further incurred the wrath of the Boston elite (who were supporters of Federalism) when he allied himself with the populist democracy espoused by Jefferson. His political enemies, consisting of a coalition of physicians from Harvard, the clergy, and others from the Boston area, arranged for his dismissal from the Harvard Medical School. In 1820, the vaccine law was repealed, and in 1822, Dr. James Smith, the federal agent for the distribution of cowpox, was dismissed from office. Politics was favored over good public health policy, and the result was that by 1840, epidemics and deaths from smallpox once again increased in the United States.

Compulsory vaccination never became a federal policy in the United States, and communities were free to vaccinate or not. However, with the expansion of elementary and secondary education after the Civil War, and eventually the compulsory attendance of children in school, the most effective way of enforcing vaccination was for public schools to require a vaccination scar on the arm before any child could attend the school.

This should have solved the problem but it did not. Until the 1900s, there were antivaccination societies, whose members believed the practice of smallpox vaccination to be dangerous, ineffective, and a violation of their civil liberties. The strength of the movement was clearly demonstrated by the "Milwaukee Riots" in 1890. The city of Milwaukee was home to an increasing number of German and Polish immigrants, who worked in the factories and lived in the poorer sections. During a smallpox epidemic in 1894, the public health authorities pursued a strict policy of enforcement, in some cases actually removing children suspected of being infected from their homes and placing them in the city's isolation hospital. The immigrant community believed not only that their civil rights were being violated but also that they were being discriminated because they were poor immigrants. Some believed that they were being poisoned, and others held to the belief that home care was better than hospitalization. They asked that if the city's smallpox hospital was so safe, why had it been built in the part of the city where the poorest people lived. When the health department ambulance came to take a child suspected of having the disease, the people threw stones and scalding water at the ambulance horses and the guards were threatened with baseball bats, potato mashers, clubs, bed slats, and butcher knives. In response, the ambulances withdrew. By the end of the epidemic, the health commissioner had been impeached. The health department also lost its police powers, which it never regained.

The vaccination controversy did not end in Milwaukee. In May 1901, there was an outbreak of smallpox in Boston. The fatality rate was 17%; there were ~1500 cases in a population of

~550,000. By the fall, the Boston Board of Health took steps to control the epidemic: a free and voluntary vaccination program was started. By December, 400,000 people had been vaccinated, but outbreaks of smallpox continued to be reported. The Board of Health established "virus squads" with the following orders: all inhabitants of the city must be vaccinated or revaccinated; there will be a house-to-house program of vaccination; and vaccine should be administered to all. Persons who refused vaccination were fined $5 or sentenced to 15 days in jail. As is frequently the case, the disenfranchised of society were blamed for spreading the disease; in this instance, it was the homeless. Virus squads were dispatched to cheap rooming houses to forcibly restrain and vaccinate the occupants. The opponents of smallpox vaccination (Anti-Compulsory Vaccination League) swung into action, complaining that vaccination was a violation of civil liberties. By January 1902, legislation was proposed to repeal the state's compulsory vaccination law. This led to a landmark legal case on the constitutionality of compulsory vaccination (*Johnson v. Massachusetts*). In 1905, the U.S. Supreme Court voted 7–2 in favor of the state, ruling that although the state could not pass laws regarding vaccination in order to protect an individual, it could do so to protect the public in the case of a dangerous communicable disease. The Boston epidemic of 1901–1903 illustrates the benefits of educating the public about the benefits of vaccination and the value of having a public debate on the pros and cons of public health policies.

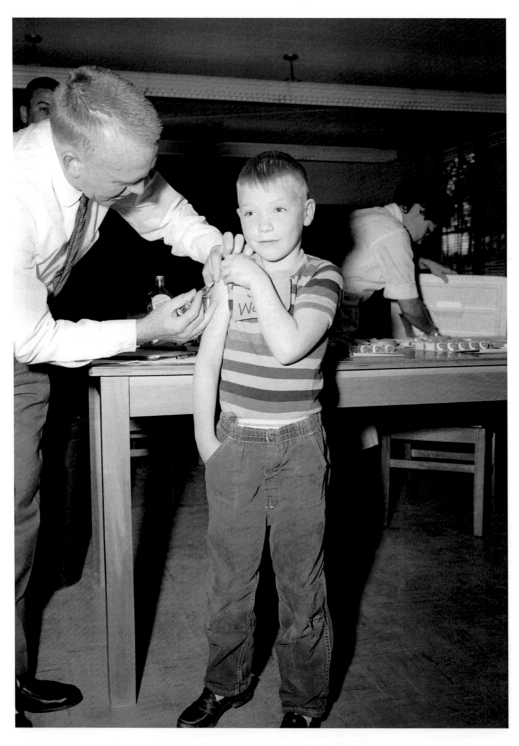

Vaccination against measles. (Courtesy of the CDC Public Health Image Library.)

Vaccines to Combat Infectious Diseases

Being immune means that the body is able to specifically and successfully react to a foreign material. To be effective, the immune response must be able to distinguish foreign substances, that is, "not self" from "self." It must be able to remember a previous encounter; that is, there must be memory. In addition, it must be economical; that is, the immune substances should not be produced all the time, but they should be turned on and off, as needed. Moreover, it must be specific for a specific antigen. Immunization, when successful, satisfies all of these properties. Indeed, it is difficult to underestimate the contribution that vaccines have made to our well-being. If there were no childhood immunizations against diphtheria, pertussis, measles, mumps, and rubella, as well as no protection afforded by vaccines against tetanus, polio, influenza, and chicken pox, childhood death rates would probably hover in the range of 20%–50%. Indeed, in those countries where vaccination is not practiced, the death rates of infants and young children remain at that level.

From One to Another: Transmission

For an infection to persist in a population, each infected individual, on average, must transmit the infection to at least one other individual. The number of individuals that each infected person infects at the beginning of an epidemic is given by R_0; this is the basic reproductive ratio of the disease, or more simply, it is the multiplier of the disease. The multiplier helps to predict how fast a disease will spread through the population.

The value for R_0 can be visualized by considering the children's playground game of "Touch Tag." In this game, one person is chosen to be "it" and the objective of the game is for that player to touch another, who, in turn, also becomes "it." From then on, each person touched helps tag others. If no other player is tagged, the game is over, but if more than one other player becomes "it," then the number of "touch taggers" multiplies. Thus, if the infected individual (it) successfully touches another (transmits), then the number of diseased individuals (touch taggers) multiplies. In this example, the value for R_0 is the number of touch taggers that result from being in contact with "it."

The longer a person is infectious and the greater the number of contacts that the infectious individual has with those who are uninfected, the greater the value of R_0 and the faster the spread of the disease. An increase in the population size or in the rate of transmission increases R_0, whereas an increase in the mortality of the infectious agent or a decrease in transmission reduces the spread of disease in a population. Thus, a change that increases the value of R_0 tends to increase the proportion of hosts infected (prevalence) as well as the burden (incidence) of a disease. Usually, as the size of the host population increases, so does the disease prevalence and incidence.

If the value of R_0 is greater than 1, then the "seeds" of the infection (i.e., the transmission stages) will lead to an ever-expanding spread of the disease—an epidemic—but, in time, as the pool of susceptible individuals is consumed (like fuel in a fire), the epidemic may eventually burn itself out, leaving the population to await a slow replenishment of new susceptible hosts (providing additional fuel) through birth or immigration. Then, a new epidemic may be triggered by the introduction of a new parasite or mutation, or there may be a slow oscillation in the number of infections, eventually leading to a persistent low level of disease. However, if R_0 is less than 1, then each infection produces less than one transmission stage and the disease cannot establish itself.

Epidemiologists know that host population density is critical in determining whether a parasite, be it a bacterium, a virus, or a protozoan, such as that of malaria, can become established and persist. The threshold value for disease establishment can be obtained by finding the population density for which $R_0 = 1$. In general, the size of the population needed to maintain an infection varies inversely with the transmission efficiency and directly with the death rate (virulence). Thus, virulent parasites, that is, those causing an increased number of deaths, require larger populations to be sustained, whereas parasites with reduced virulence may persist in smaller populations.

The spread of infection from an infected individual through the community can be thought of as a process of diffusion, where the motions of the individuals are random and movement is from a higher concentration to a lower one. Therefore, factors affecting the spread of an infection include the size of population, those communal activities that serve to bring the susceptible individuals in contact with infectious individuals, the countermeasures used (e.g., quarantine, hospitalization, and immunization), and seasonal patterns. For example, in northern temperate climes, measles is spread most frequently in the winter months, because people tend to be confined indoors. Other uncertainties in predictability may involve changes in travel patterns with contact and risk increased. Sociological changes may also affect the spread of disease—children in school may influence the spread of measles. Quarantine of infected individuals has also been used as a control measure. Generally speaking, quarantine is ineffective, and more often than not, it is put in place to reassure the concerned citizens that steps at control are being taken. However, as noted above, there are other interventions that do affect the spread of disease by reducing the number of susceptible individuals. One of the more effective measures is immunization.

To block parasite transmission, a sufficient number of individuals in the population must be immunized, such that the value of R_0 is less than 1. For measles, R_0 is approximately 10. With this multiplier, measles will spread explosively; indeed, with multiplication every two weeks and without any effective control (such as immunization), millions could become infected in a few months. It has been estimated that to eliminate measles (and whooping cough), ~95% of children under the age of 2 years must be immunized; for mumps and rubella to be eliminated, the percentages of children that need to be immunized are 90% and 85%, respectively. Therefore, if transmission is intense (R_0 is a large number), mass immunization must take place at the earliest age feasible, and the later the average age of vaccination, the less likely it is that transmission will be blocked. For disease elimination, not everyone in the population needs to be immunized, but it is necessary to reduce the number of susceptible individuals below a critical point (called herd immunity).

Diphtheria

Diphtheria is a horrible disease. The 7-day incubation period is followed by fever, malaise, a sore throat, hoarseness, wheezing, a blood-tinged nasal discharge, nausea, vomiting, inflamed tonsils, and headache. The infected child is quite pale and has a sticky gray-brown membrane (made of white cells, red cells, dead epithelial cells, and clots) that adheres to the palate, throat, windpipe, and voice box. This membrane impairs the child's breathing. If disturbed, it can bleed and break off and block the airways of the upper respiratory tract. Speech becomes thickened, and the breath is foul smelling. Within 2 weeks, this upper respiratory infection can spread to the heart, leading to inflammation of the heart, heart failure, circulatory collapse, and paralysis. During an epidemic, up to 50% of untreated children die, but more commonly, the death rate is 20%.

In 1883, two German scientists, Edwin Klebs and Friedrich Loeffler, working in Robert Koch's laboratory in Berlin, discovered the microbe of diphtheria in throat swabs taken from an infected child; they were able to grow the bacterium in pure culture and named it *Corynebacterium diphtheriae*. Then, in 1888, Emile Roux and Alexandre Yersin of the Pasteur Institute in Paris found that the broth in which the *Corynebacterium* bacteria were growing contained a substance that produced all the symptoms of diphtheria when injected into a rabbit, and at high-enough doses, it killed the rabbit. This killing agent was a poisonous toxin. In humans, it is this toxin, that leads to heart muscle degeneration; neuritis; paralysis; and hemorrhages in the kidneys, adrenal glands, and liver. The toxin is produced because the diphtheria bacterium itself is infected with a virus. In order for the toxin to be synthesized, iron molecules must be present in its surroundings—body tissues or culture medium. Today, diphtheria is less of a problem because of childhood immunization (the DaPT "shot"), or if the infection is acquired, it can be treated with antibiotics.

Tetanus

In 1885, Arthur Nicolaier discovered the tetanus bacillus *Clostridium tetani*, and in 1889, Shibasaburo Kitasato in Robert Koch's Berlin laboratory was able to grow it in pure culture and found a powerful toxin in the culture fluid. Then, in 1890, two critical discoveries related to protective immunity were made: Emil von Behring and Shibasaburo Kitasato, both working with Koch, found that if small amounts of tetanus toxin were injected into a rabbit, the rabbit's serum contained a substance that would protect another animal from a subsequent lethal dose of toxin. (Serum is the straw-colored liquid that remains when blood is allowed to clot and then the clotted material is removed.) They called the serum that was able to neutralize the toxin "immune serum." Emil von Behring found that the same situation also occurred with diphtheria toxin, a toxin so powerful that a single molecule can kill a cell. In effect, by toxin inoculation, the rabbit had been protected against a disease. This process was called immunization, although in Jenner's honor, any immunization is called a vaccination. Further, the immune protection seen by von Behring and Kitasato could be transferred from one animal to another by injection of this "immune serum." In other words, it was possible to have passive immunity, that is, immunity could be borrowed from another animal that had been immunized with the active foreign substance when its serum was transferred (by injection) to a nonimmunized animal. (Passive immunization or "immunity on loan" can be lifesaving in the case of toxins such as bee, spider, and snake venoms or tetanus, since it is possible to counteract the deadly effects of the poison by injection of serum containing the appropriate antitoxin.) The implications of the research on "immune serums" for treating human disease were quite obvious to von Behring and others in the Berlin laboratory. In addition, without waiting for the identification of the active ingredient in "immune serum," the German government began to support the construction of factories to produce different kinds of "immune sera" against a wide range of bacterial toxins. Clinical trials were organized first in Germany and then in Europe and the United States.

Von Behring and Kitasato called the substance in "immune serum" that neutralized these toxins "antitoxins," meaning "against the poison." Treatment of diphtheria cases with antitoxin serum (or more simply antiserum produced in horses) was so effective that the death rate dropped precipitously. In some cases, mortality was halved. For his work on antitoxin therapy, von Behring received the first Nobel Prize in 1901. Initially, it was believed that all bacterial infections could be treated with antitoxin therapy, but unfortunately, only the bacteria that cause diphtheria and tetanus and a few others produce and secrete toxins, so the therapeutic application of "immune serum" against most bacterial infections remains limited.

Antitoxins and Toxoids

An antitoxin is one of many kinds of active materials found in "immune serum" after a foreign substance is injected into humans, rabbits, horses, chickens, mice, guinea pigs, monkeys, or rats; the general term used for the active material is antibody. Simply put, antitoxin is one kind of antibody. At about this time, Paul Ehrlich (1854–1915) entered the scene in Koch's laboratory. Ehrlich established the principle that there was a quantitative (1:1) relationship between the amount of toxin neutralized by a specific amount of antitoxin. The potency or strength of an antibody is called its titer and is reflected by the degree of dilution that must be made to have a specific amount of antigen that will bind to an equal amount of antibody. (The higher the titer, the greater the potency of the serum, or put another way, a high-titer "immune serum" can be diluted to a greater degree to achieve the same neutralizing effect as would be achieved by a weaker "immune serum.") Ehrlich's principle of titer was not only of theoretical interest, but it also enabled antitoxins (and other therapeutic biologics such as insulin and other hormones) to be provided in standardized amounts and strengths. For his work on immunity, Ehrlich shared the Nobel Prize with Metchnikoff in 1908. Ehrlich also found that although the diphtheria toxin lost its poisoning capacity with storage, it still retained its ability to induce "immune serum." He called this altered toxin (usually produced now by treatment with formalin) "toxoid." Even today, this is the standard means of inducing immunity to diphtheria, pertussis, and tetanus. Toxoids made from diphtheria, pertussis, and tetanus toxins are used as the DPT vaccine. The Schick test for susceptibility to diphtheria also involves toxoid: a small amount of diphtheria toxoid is injected just under the skin surface; failure to react (by an absence of a red swelling—inflammation—at the site of injection) indicates a lack of protective immunity. Individuals that are immunosuppressed, such as acquired immune deficiency syndrome (AIDS) patients, those on anticancer chemotherapy, and those receiving high doses of steroids, cannot mount an immune response and give false negatives with such tests.

Ehrlich went on to study other toxins and found that the serum made against one toxin did not protect against another toxin. Thus, the antitoxin made by the animal was specific. Moreover, since this specificity was true for almost any foreign substance, he concluded that antibodies were specific for the foreign material against which they were produced. The foreign substances that were antibody generators, such as a toxin and a toxoid, were called antigens. Antigens are any material that is foreign to the body, for example, a bacterium or its toxin, a virus such as the human immunodeficiency virus (HIV), a piece of tissue from another individual, and a foreign chemical substance such as a protein, a nucleic acid, or a polysaccharide (a carbohydrate polymer).

Why Do Antibodies Work?

Paul Ehrlich also formulated a theory of how antigen and antibody interact with one another. It was called the "lock-and-key" theory. He visualized it in the following way: the tumblers of the lock were the antigen, and the teeth of the key were the antibody. One specific key would work to open only one particular lock. Another way of looking at the interaction of antigen with antibody is the way a tailor-made glove is fitted to the hand. One such glove will fit only one specifically shaped hand. Similarly, only when the fit is perfect can the antigen combine with the antibody. Yet, despite the fact that antigens are large molecules, only a small surface region is needed to determine the binding of an antibody to it. These antigenic determinants are called epitopes, and they can be thought of as a small patch on the surface of the much larger molecule of antigen. Earlier, when vaccination was discussed (see p. 83), it was said, without any explanation, that protection occurred because the antibodies produced against *Variolae vaccinae* were able to neutralize the *V. major* virus. Simply stated, it is because the epitopes of the cowpox antigens are so similar to those of smallpox that when the body produces antibodies to cowpox antigens, these are also able to fit exactly to the smallpox epitopes (i.e., the antibodies are said to be "cross-reactive"), and *V. major* is prevented from multiplying.

Attenuation and Immunization

In 1875, Louis Pasteur attempted to induce immunity to cholera. He was able to grow the cholera bacterium that causes death in chickens and to reproduce the disease when healthy chickens were injected with it. The story is told that he placed his cultures on a shelf in his laboratory, exposed to the air, and went on summer vacation; when he returned to his laboratory, he injected chickens with an old culture. The chickens became ill, but to Pasteur's surprise, they recovered. Pasteur then grew a fresh culture of the bacterium, with the intention of injecting another batch of chickens, but the story goes that he was low on chickens and used those he had previously injected. Again, he was surprised to find that the chickens did not die from cholera; they were protected from the disease. To the prepared mind of Pasteur, he recognized that aging and exposure to air (and possibly the heat of summer) had weakened the virulence of the bacterium and that such an attenuated or weakened strain might be used to protect against disease. In honor of Jenner's work a century earlier, he called the attenuated strain a vaccine. Pasteur extended these findings to other diseases, and in 1881, he was able to protect sheep against anthrax. In 1885, he administered the first vaccine (prepared from dried spinal cords from rabies-infected animals) to a human, a young boy, Joseph Meister, who had been bitten by a rabid dog. The boy lived. Pasteur proved that vaccination worked.

After Pasteur's success in creating standardized and reproducible vaccines at will, the next major step came in 1886 from the United States. Theobald Smith and Edmund Salmon published a report on their development of a heat-killed cholera vaccine that immunized and protected pigeons. Two years later, these two investigators claimed priority for having prepared the first killed vaccine. Although their work appeared in print 16 months before a publication by Chamberland and Roux (working in the laboratory of Louis Pasteur) and with identical results, the fame and prestige of Pasteur was so widespread that the claim of "first killed vaccine" by Smith and Salmon was lost in the aura and accorded those who worked in Pasteur's Institute in Paris. During the period 1870–1890, killed vaccines for typhoid fever, plague, and cholera were developed, and by the turn of the century, there were two human attenuated virus vaccines (smallpox and rabies) and three killed ones.

Influenza

The 1918 outbreak of influenza remains one of the world's greatest public health disasters. Some have called it the twentieth century's weapon of mass destruction. It killed more people than the Nazis and far more than the two atomic bombs dropped on Japan. Before it faded away, it affected 500 million people worldwide and 20–40 million people perished. This epidemic killed more people in a single year than the Black Death did in 100 years. In 24 weeks, influenza destroyed the lives of more people than AIDS had in 24 years. As with AIDS, it killed those in the prime of their life: young men and women in their 20s and 30s. During its 2-year course, more people died from the flu than from any other disease in recorded history.

The 1918–1920 flu epidemic brought more than human deaths: civilian populations were thrown into panic; public health measures were ineffectual or misleading; there was a government-inspired campaign of disinformation; and people began to lose faith in the medical profession. Because the flu pandemic killed more than twice the number of people who died on the battlefields of World War I, it hastened the armistice that ended the "Great War" in Europe.

Where did this global killer come from? Epidemiologic evidence suggests that the outbreak was due to a novel form of the influenza virus that arose among the 60,000 soldiers billeted in the army camps of Kansas. The barracks and tents were overflowing with men, and the lack of adequate heating and of warm clothing forced the recruits to huddle together around small stoves. Under these conditions, they shared both the breathable air and the virus that it carried. By mid-1918, flu-infected soldiers carried the disease by rail to army and navy centers on the East Coast and the South. Then, the flu moved inland to the Midwest and onward to the Pacific states. In its transit across America, cases of influenza began to appear among the civilian population.

People in cities such as Philadelphia, New York, Boston, and New Orleans began to ask: What should I do? How long will this plague last? To minimize panic, the health authorities and newspapers claimed it was "la grippe," and there was little cause for alarm. This was an outright lie. However, as the numbers of civilian cases kept increasing and when there were hundreds of thousands sick and hundreds of deaths each day, it was clear that this flu season was nothing to be sneezed at and that the peak of the epidemic had not been reached. In Philadelphia, undertakers had no place to put the bodies and there was a scarcity of coffins. Grave diggers were either too sick or too frightened to bury influenza victims. Entire families were stricken, with almost no one to care for them. There were no vaccines or drugs. None of the folk remedies were effective in stemming the spread of the virus, and the only effective treatment was good nursing. In cities, as the number of sick continued to soar, public gatherings were forbidden and gauze masks had to be worn as a public health measure. The law in San Francisco was: If you do not wear a mask, you will be fined or jailed.

Conditions in the United States were exacerbated as the country prepared to enter the war in Europe. President Woodrow Wilson's aggressive campaign to wage total war led indirectly to the nation becoming "a tinder box for epidemic disease." A massive army was mobilized, and millions of workers crowded into the factory towns and cities, where they breathed the same air and ate and drank using common utensils. With an airborne disease such as the flu, this was a prescription for disaster. The war effort also consumed the supply of practicing physicians as well as nurses, and so medical and nursing care for the civilian population deteriorated. "All this added kindling to the tinderbox."

The epidemic spread globally, moving outward in ever-enlarging waves. The hundreds of thousands U.S. soldiers that disembarked in Brest, France, carried the virus to Europe and the British Isles. Flu then moved to Africa via Sierra Leone, where the British had a major coaling center. The dockworkers who refueled the ships contracted the infection and they spread the highly contagious virus to other parts of Africa when they returned to their homes. In Samoa, a ship arrived from New Zealand, and within 3 months, over 21% of the population had died. Similar figures for deaths occurred in Tahiti and Fiji. In a few short years, the flu was worldwide and death followed in its wake.

During the epidemic, there was a rhyme to which little girls jumped rope: I had a little bird/And its name was Enza/I opened the window/And in flew Enza. This delightful singsong rhyme did not describe Enza's symptoms, some of which you have probably experienced yourself: fever, chills, sore throat, lack of energy, muscular pain, headaches, nasal congestion, and a lack of appetite. It can escalate quickly to produce bronchitis and secondary infections, including pneumonia, and can lead to heart failure and, in some cases, death.

The influenza virus is highly infectious (it has an R_0 value of 10), and it spreads from person to person by droplet infection through coughing and sneezing. Each droplet can contain between 50,000 and 500,000 virus particles. However, the virus does not begin its "island hopping" in humans. Instead, aquatic wild birds such as ducks and other waterfowl maintain the flu viruses that cause human disease. Because these aquatic birds, which carry the viruses in their intestines, do not get sick and can migrate thousands of miles, the virus can be carried across the face of the Earth before humans enter the picture. However, the flu virus found in these waterfowl is unable to replicate itself well in humans, and for it to become a human pathogen, it must first move to an intermediate host—usually domestic fowl (chickens, geese, or ducks) or pigs—that drinks water contaminated with the virus-containing feces of the wild waterfowl. The domestic fowl tend to be dead-end hosts, because some sicken and die, but pigs live long enough to serve as "virus-mixing vessels"—and here, the flu genes of bird, pig, and human can be mixed because pig cells have receptors for both bird and human viruses. It is by this coming together of virus genes that new strains of the flu are produced.

Let's consider the eight genes of the flu virus as if they were a small deck of playing cards. If the eight different cards are shuffled and two cards are removed from the pack at the same time, how many different combinations are there? The answer is 256 or 2^8. In the same way, if two different viruses infect the same cell and genes are exchanged randomly, there can be 256 different virus offsprings. This mixing tends to occur where birds, pigs, and humans live in close proximity—predominantly in China (where the 2002–2003 severe acute respiratory syndrome [SARS] outbreak also began). The 1997 Hong Kong "bird flu" was a flu virus that became virulent by acquiring genes from geese, quail, and teal, because these birds were housed together in Hong Kong poultry markets, where mixing could occur quite easily. This flu strain killed thousands of chickens, before humans were infected. Eighteen people acquired the infection from contact with the feces of infected chickens, not from other people. Fortunately, the spread to other humans was halted before person-to-person (droplet) transmission could result in a full-blown flu epidemic, because health authorities enforced the slaughter of more than a million fowl in Hong Kong's markets. Had this not occurred, one-third of the human population may have sickened and died.

What are the genes that produce the new and virulent flu strains? They principally involve two surface proteins, which are both critical to virus entry into cells, where the mixing of genes takes place. One is called hemagglutinin (H) and the other is neuraminidase (N). The job of the surface spikes of H is to act like grappling hooks to anchor the influenza virus to host cell receptors called sialic acid. After binding, the virus can enter the host cell, replicate its RNA, and produce new viruses. The emerging viruses are coated with sialic acid, the substance that enabled them to attach to the host cell in the first place. If the sialic acid were allowed to remain on the virus and on the host cell, then these new virus particles with H on their surface would be clumped together and trapped much like flies sticking on flypaper. The N, which resembles lollipops on the virus surface, allows the newly formed viruses to dissolve the sialic acid "glue"; this separates the viruses from the host cell and allows them to plow through the mucus between the cells in the airways along the respiratory tract and to move from cell to cell. The entire process—from anchoring to release—takes about 10 h, and in that time, 100,000 to 1 million viruses can be produced.

There are 15 different Hs and 9 different Ns, all of which are found in bird flu viruses. A letter and a number designate each flu strain. For example, H1N1 is the strain that caused the 1918 pandemic, and H5N1 is the 1997 Hong Kong flu strain. Flu epidemics occur when either H or N undergoes a genetic change due to a mutation in one of the virus genes. If a virus never changed its surface antigens, as is the case with measles and mumps, then the body could react with an immune response—antibody- and cell-mediated—to the foreign antigens (see p. 92) during an infection or by a vaccine and there would be long-lasting protection. Indeed, if a person encounters the same virus, the immune system, having been primed, will swiftly eliminate that virus and infection will be prevented. However, in the case of flu, there may be no immune response, because the virus changes the N and H molecules—sites where the antibodies would ordinarily bind to neutralize the virus. This mutational change of the flu virus (called antigenic drift) ensures escape from immune surveillance and allows the virus to circumvent the body's ability to defend itself. Such slight changes in antigens lead to repeated outbreaks during interpandemic years. Every 10–30 years, a more radical and dramatic change in the virus antigens (called antigenic shift) may occur, and so, approximately every quarter of a century, an influenza virus emerges that the human immune system has never encountered before. When this happens, a pandemic can occur. There have been flu pandemics in both 1957 (Asian flu) and 1968 (Hong Kong flu), and flu scares in 1976 (swine flu) and 1977 (Russian flu). In 2004, the H5N1 strain resulted in 27 cases and 20 deaths in Vietnam and 16 cases and 11 deaths in Thailand. Still to be explained is why the 1918 outbreak of influenza was so pathogenic. It might have had something to do with an exaggerated innate inflammatory immune response with release of lymphokines such as tumor necrosis factor; this could lead to a toxic-shock-like syndrome, including fever, chills, vomiting, and headache, and ultimately could result in death. Alternatively, the H1N1 virus, which led to the 1918–1920 pandemic, had a very different kind of H, one more closely related to that of a swine flu, not one from birds, and for this, there was little in the way of immune recognition.

Globally, influenza remains an important contagious disease, with 20% of children and 5% of adults developing symptoms. The death rate can be as high as 5%. Each fall, we are encouraged to get a "flu shot," so that when the flu season arrives in winter, we are protected. However, flu shots can protect only against targeted or known strains—those whose antigenic type has been determined by scientists at the World Health Organization (WHO)—not from unexpected or unidentified types. The flu vaccines used today do protect against a known type and they do not produce disease, because after the viruses have been grown in chicken embryos, they are purified and inactivated. Live flu viruses (unlike the viruses of polio and *V. vaccinae* that are used in immunizations) are not used in vaccines, and so, immunity, not infection, results. Although some vaccines consist of only a portion of the virus, such as H or N antigens, and these serve to activate the immune response, weakened live-virus vaccines may give a stronger and longer-lasting protection. However, for these vaccines, it is critical that a return to virulence does not occur. Although these vaccines may be administered as a nasal spray, avoiding the pain of inoculation, they may also generate disease symptoms. Flu outbreaks will continue to plague humankind as long as there is "viral mixing," and in those at high risk (i.e., those over 50 years of age, the very young, the chronically ill, and the immunosuppressed), it may be impossible to prevent infection under any circumstances. In addition, those infected with the flu are at greater risk for pneumococcal pneumonia (a bacterial disease), which annually kills thousands of elderly people in the United States. Flu combined with pneumonia can result in skyrocketing mortality.

For a disease such as the flu, with its high R_0 value (see p. 88), quarantining those who show symptoms is not enough to bring the average number of new infections caused by each case to below one—the level necessary for an epidemic to go into decline. Therefore, other measures, such as treatment and immunization, have to be employed. Flu treatments involve antiviral drugs (zanamivir and oseltamivir) that block the synthesis of the neuraminidase. When administered early enough, the complete virus cannot be released. In the case of other antiflu drugs (amantidine and rimantidine), the viruses quickly acquire resistance, and so, these are less effective in preventing infection or in reducing the severity and duration of symptoms.

Another pandemic of influenza is inevitable. Modern means of transportation—especially jet airplanes—ensure that the virus can be spread across the globe in a matter of hours or within a day in an infected traveler. Surveillance may alert us of the possibility, and drugs may reduce the severity of illness, but neither of the two can guarantee when, where, or how lethal the next pandemic will be. What is predictable is that it will impact our lives: hospital facilities will be overwhelmed because medical personnel will also become sick; vaccine production will be slower because many of the personnel in pharmaceutical companies would be too ill to work; and reserves of vaccines and drugs will soon be depleted, leaving most people vulnerable to infection. There will be social and economic disruptions. How can we prepare for a "future shock" such as that of 1918–1920 pandemic? Stockpiling of anti-infective drugs, promoting research on infectious diseases, and increasing the methods of surveillance may all help blunt the effects, but these measures cannot eliminate them.

Childhood Diseases and Vaccines

Measles, mumps, whooping cough, and chicken pox pass from one individual to another without any intermediary; recovery from a single exposure produces lifelong immunity. As a consequence, these diseases commonly afflict children, and for that reason, they are called "childhood diseases."

Measles, Mumps, and Rubella

In the year 2015, for some, Disneyland was not the happiest place on Earth. It was in January of that year that a single measles-infected individual was able to spread the disease to 145 people in the United States and to a dozen others in Canada and Mexico. Patient zero in the 3-month-old Disneyland outbreak was probably exposed to measles overseas, and unknowingly, while contagious, he visited the park. (The measles strain in the Disneyland outbreak was found to be identical to the one that spread through the Philippines in 2014, where it sickened about 50,000 and killed 110 people. It is likely that patient zero acquired the virus there.)

Measles spreads from person to person by sneezing and coughing; the virus particles are hardy and can survive as long as 2 h on doorknobs, handrails, elevator buttons, and even in air. For the first 10–14 days after infection, there are no signs or symptoms. A mild to moderate fever, often accompanied by a persistent cough, runny nose, inflamed eyes (conjunctivitis), and sore throat follows. This relatively mild illness may last 2 or 3 days. Over the next few days, the rash spreads down the arms and trunk, then over the thighs, lower legs, and feet. At the same time, fever rises sharply, often as high as 104°F–105.8°F (40°C–41°C). The rash gradually recedes, and usually, lifelong immunity follows recovery. Complications occur in about 30% of cases and may include diarrhea, blindness, inflammation of the brain, and pneumonia. Between 1912 and 1916, there were 5300 measles deaths per year in the United States; however, all that changed in 1968 with the introduction of the measles vaccine. In the United States, measles was declared eliminated in 2000. So, what underlies the Disneyland outbreak?

On average, every measles-infected person is able to spread the disease to 10 other people. Analysis of the outbreak shows that those infected were unvaccinated. Indeed, it is estimated that in order to establish herd immunity, between 96% and 99% of the population must be vaccinated. Disease experts who analyzed the Disneyland outbreak of measles have calculated that the number of vaccinated individuals might have been as low as 50%. The outbreak that began in California was a reflection of the antivaccination movement, where some parents believe in the false claim that the vaccine for measles caused an increase in autism. (The "evidence" for this was based on just 12 children and has been thoroughly discredited by massive studies involving half a million children in Denmark and 2 million children in Sweden.) Then too, some parents believe that their children are being immunized too often and with too much vaccine because pharmaceutical companies are recklessly promoting vaccination in pursuit of profit. Other parents contend that the vaccine is in itself dangerous. It is not as evidenced in Orange County, where Disneyland is located: the outbreak sickened 35 people, including 14 children. Moreover, although a measles vaccine has been available worldwide for decades, according to the WHO, about 400 people a day died in 2013.

The response to the outbreak at Disneyland prompted the California Senate to pass a bill, SB 277, which required almost all California schoolchildren to be fully vaccinated in order to attend public or private school, regardless of their parents' personal or religious beliefs. When signing the bill, Governor Edmund G. (Jerry) Brown wrote: "While it is true that no medical intervention is without risk, the evidence shows that immunization powerfully benefits and protects the community."

The measles virus is unlike the bacterial diseases diphtheria and tetanus in that it cannot be grown in anything but living cells. Making the measles vaccine first required growing the virus in the laboratory in tissue culture—glass vessels seeded with living cells. It was first developed in 1928 by a Canadian husband and wife team, Hugh and Mary Maitland, working in Manchester, United Kingdom. Using minced kidney, serum, glucose, and mineral salts, they were able to grow smallpox virus and other viruses. This was a great breakthrough, since until their study, all work with viruses had to be conducted using laboratory animals (mice or monkeys). In 1931, Ernest Goodpasture and Alice Woodruff developed another technique for growing viruses: chick embryos. This too has been used ever since for the manufacture of some vaccines.

In January 1954, John Enders at the Boston Children's Hospital in Boston assigned to a young MD, Thomas Peebles, the task of capturing the measles virus and then propagating it in tissue culture for future use as a vaccine. When it was learned that there was an outbreak of measles at the all-boys Fay School, Peebles went to the school and convinced the principal to allow him to collect blood samples. On February 8, 1954, Peebles collected blood from a 13-year-old student, David Edmonston, who was clearly suffering from measles—he was nauseous, had a fever, and had a telltale red rash. Edmonston's blood was added to a culture of human kidney cells, and within a few days, it was clear that the measles virus was growing and killing the cells. Following the work of Pasteur on attenuation, it was hoped that by forcing the virus to grow in a hodgepodge of different kinds of cells, the measles virus would become sufficiently weakened so that it could serve as a potential vaccine. Over many months, the measles virus was passed serially: 24 times in human kidney cells, 28 times in cultures of cells from the placenta, and six times in minced chick embryos. Believing that the virus had been sufficiently attenuated, in 1958, the putative measles vaccine was tested in 11 mentally retarded children at the Fernald School, where measles occurred annually, with a high morbidity and mortality. In the trial, all the children developed antibodies to the virus but some experienced fever and had a mild rash. Clearly, the measles virus had not been sufficiently attenuated. In 1958–1959, the chick-embryo-cell-cultured measles virus was injected into monkeys; they developed antibodies but no illness, and no virus was detectable in the blood. This attenuated virus was then administered to immune adults, with no ill effects. In February 1960, this vaccine was tested at the Willowbrook State School in mentally retarded children. Twenty-three children received the vaccine, and 23 other children received nothing. Six weeks later, there was an outbreak of measles at the Willowbrook School, with hundreds of children becoming infected; however, none of those vaccinated came down with measles. Many, but not all, of the unvaccinated also became infected. Again, there was a high rate of side effects, including fever and rash. It was suggested that the toxic effects of the vaccine might be ameliorated by the addition of a small amount of gamma globulin along with the vaccine. This was tested at a woman's prison in New Jersey, Clinton Farms for Women, where there was also a nursery full of their babies. Six of the infants were given the vaccine in one arm and the gamma globulin in the other. None of the babies had high fevers, and only one baby had a mild rash. Subsequent studies were carried out with susceptible children chosen on the basis of an absence of antibodies to the measles virus and after parental permission. The success of these studies led to further trials among home-dwelling children in five U.S. cities. The Edmonston measles virus was further attenuated by Maurice Hilleman at Merck by passage 40 more times in chick embryos growing at lower temperatures; this attenuated strain, called Moraten (for more-attenuated enders), which did not require gamma globulin, was licensed in 1968 and has been used throughout the world. Before the development of the measles vaccine, 7–8 million children around the world died from measles every year. In the United States, between 1968 and 2006, hundreds of millions of doses were given, and the number of measles cases decreased from 4 million to fewer than 50. It is estimated by the WHO that between 2000 and 2003, the measles vaccine had prevented 15.6 million deaths worldwide.

Mumps

In the 1960s, the virus that causes mumps infected a million people, mostly children in the United States. The mumps virus attacks the salivary glands in the front of the ears, causing children to look like chipmunks. Sometimes, the virus infects the brain and the spinal cord, causing meningitis, paralysis, and deafness. In males, it might also infect the testes, resulting in sterility; in pregnant females, it could result in birth defects and fetal death. If the virus attacks the pancreas, it could cause diabetes. The spherical mumps virus contains a single strand of RNA that codes for nine proteins. The HN (hemagglutinin-neuraminidase) proteins are essential for viral attachment to the host cell and release from the cell, and the fusion (F) protein is essential for viral penetration into the host cell. The nucleocapsid protein (NP) encloses the RNA (ribonucleoprotein, RNP); there is a matrix protein, M; and the nonstructural NS 1 and NS 2 proteins have no known function. The V protein is involved in the pathogenic effects.

On March 23, 1963, Maurice Hilleman's 5-year-old daughter Jeryl Lynn came down with mumps. Hilleman, the then Director of Virus and Cell Biology at Merck Research Laboratories, stroked the back of her throat with a swab and placed it in a vial of nutrient broth, which was then used to inoculate an egg containing a chick embryo. The virus grew, and after passing it several times in embryos, Hilleman placed it into a flask containing chick embryo brain cells. This procedure was repeated over and over again to attenuate the mumps virus. To test the protective capacity of the attenuated virus, Hilleman (who was a PhD, not an MD) contacted the pediatrician Robert Weibel and Joseph Stokes, chairman of pediatrics at Children's Hospital, in Philadelphia. In June 1965, Weibel tested the putative vaccine in 16 mentally retarded children at the Trendler School in Bristol, Pennsylvania, where Weibel's own son with Down's syndrome was a resident. The vaccine was found to be safe, and antibodies developed against the virus. Encouraged by the findings, in August, Weibel carried out additional tests in 60 severely retarded children. The results were the same—antibodies to the mumps virus and no illness. To prove the vaccine was protective, Stokes and Weibel recruited parents of kindergartners to participate in a more extensive clinical trial. Four-hundred children were enrolled in the study, with 200 receiving the vaccine (the attenuated Jeryl Lynn strain of the mumps virus) and 200 receiving no vaccine. Sixty-three children came down with mumps, with only 2 of the 63 having been vaccinated. On March 30, 1967, 4 years after Hilleman's daughter came down with mumps, the vaccine was licensed. Since then, hundreds of million doses have been distributed in the United States.

Rubella

A German physician, originally described rubella, or German measles, at the end of the eighteenth century. For years, the skin rash of rubella was confused with measles and scarlet fever, but in 1881, a consensus was finally reached that rubella was a specific viral disease. Rubella begins as an upper respiratory infection, with virus multiplication in the nasopharynx and then in the lymph nodes. After 14 to 21 days, there is a fine rash that begins on the face and spreads out over the entire body. After the rash subsides, there may be complications such as arthritis, encephalitis, and platelet depression. The most serious complication is when the infection is acquired during the first trimester of pregnancy, where a woman has a better than a two in three chances of giving birth to an infant that is deaf, blind, or mentally retarded. A famous case was that of the movie star Gene Tierney. When she gave birth to a severely deformed child 8 months after being kissed during a USO tour by a fan with German measles, the retarded infant was placed in an institution and Tierney had to be hospitalized with severe depression.

In 1961, during a rubella outbreak at Ft. Dix, two Walter Reed Army Institute physicians took throat washings from some hospitalized recruits and placed these into cultures of African green monkey kidney (AGMK) cells. The virus that grew was named the Parkman strain after one of the physicians. Over 77 passages in AGMK cells were needed to attenuate the virus, so it could be tested in human subjects. This high-passage Parkman strain (named HPV-77) was given to Maurice Hilleman at Merck, who then adapted it to duck embryo cultures. However, before Hilleman could get the attenuated rubella strain into production as a vaccine, Mary Lasker, the widow of the advertising millionaire Albert Lasker and a philanthropist involved in health funding, paid him a visit. Lasker noted that Parkman, then at the National Institutes of Health, was already working on a rubella vaccine and she warned Hilleman: "You should get together and make one vaccine or else you'll have trouble getting yours licensed." Hillman agreed and modified the Parkman HPV-77 vaccine, and by 1969, the rubella vaccine was ready for marketing in the United States.

A competitor to Hilleman was Stanley Plotkin. Plotkin had grown up in the Bronx. He went to the Bronx High School of Science, a highly competitive school for gifted students, and was enamored by reading Paul deKriuf's *Microbe Hunters*. He attended New York University (NYU) and the Downstate Medical Center in Brooklyn, both on full scholarships. Plotkin then worked at the Centers for Disease Control and Prevention and then at the Wistar Institute, Philadelphia. In 1963, during an outbreak of rubella, there were thousands of damaged fetuses from women who were anxious about their pregnancies and were having therapeutic abortions. In 1964, Plotkin obtained an aborted fetus from an infected mother. Because this was the 27th aborted fetus that Plotkin had received and because he was able to isolate rubella from the third organ he tested—the kidney—Plotkin called his virus Rubella Abortus 27/3 (RA27/3). Plotkin attenuated RA27/3 by growing it for 25 consecutive passages at low temperature in fetal cells named WI-38 (for Wistar 38). It was immunogenic when given by the intranasal route. Clinical trials were carried out between 1967 and 1969, and it was licensed in the United Kingdom in 1970. There was little interest in a rubella vaccine in the United States, until 1978, when Plotkin received a phone call from Hilleman, suggesting that the latter's vaccine (HPV-77) be replaced with Plotkin's RA27/3. This was agreed to and Merck carried out large-scale clinical trials, and the vaccine was licensed in the United States in 1979. Today, this is the only rubella vaccine used throughout the world.

The impact of the rubella vaccine can be seen in a comparison of the number of cases. During 1964–1965, there were 12,500,000 cases of rubella in the United States, with 20,000 cases of congenital rubella syndrome (CRS). Between 1970 and 1979, there were 1064 cases of CRS or 106 cases per year, whereas from 1980 to 1985, there were 20 CRS cases per year. Since that time, rubella has been eliminated from the United States.

Whooping Cough

Whooping cough was first described in 1576, and by 1678, its name "pertussis" from the Latin, meaning "intensive cough," was in common use in England. In 1906, Jules Bordet, the director of the Pasteur Institute in Brussels, and his brother-in-law, Octave Gengou, isolated from a child with whooping cough the bacterium that causes the disease and named it *Bordetella pertussis*. Pertussis is a highly contagious disease. Once infected, it takes about 7–10 days for signs and symptoms to appear, although it can sometimes take longer. At first, it resembles a common cold with a runny nose, nasal congestion, watery eyes, and fever, but after a week or two, thick mucus accumulates inside the airways, causing uncontrollable fits of coughing. Following a fit of coughing, a high-pitched whoop sound or gasp may occur as the person breathes in. The coughing may last for 10 or more weeks, hence the phrase "100-day cough." A person may cough so hard that he or she may vomit, break ribs, or become extremely fatigued. The disease spreads easily through coughs and sneezes, and people are infectious to others from the start of symptoms, until about three weeks into the coughing fits. It is estimated that 16 million people worldwide are infected per year. Most cases occur in the developing countries, and people of all ages may be affected. In 2013, whooping cough resulted in

61,000 deaths. Pertussis is fatal in an estimated 1.6% of hospitalized U.S. infants under 1 year of age. First-year infants are also more likely to develop complications such as pneumonia (20%), encephalopathy (0.3%), seizures (1%), failure to thrive, and death (1%)—perhaps owing to the ability of the bacterium to suppress the immune system.

Efforts to develop an inactivated whole-cell whooping cough vaccine began soon after Bordet and Gengou published their method of culturing *B. pertussis.* In 1925, the director of the Danish State Serum Institute, Thorvald Madsen, used a killed whole-cell vaccine to control outbreaks in the Faroe Islands in the North Sea. In 1942, Grace Eldering and Pearl Kendrick, working in the State of Michigan Health Department, combined their improved killed whole-cell pertussis vaccine with diphtheria and tetanus toxoids to generate the first DTP combination vaccine. To minimize the frequent side effects caused by the pertussis component, in 1981, Yuji Sato in Japan developed an acellular pertussis (aP) vaccine consisting of a purified filamentous hemagglutinin that is surface localized as well as secreted by *B. pertussis*; the hemagglutinin mediates adherence to epithelial cells and is essential for bacterial colonization of the trachea. First introduced in Japan, the vaccine was approved in the United States in 1992 for use in the combination DTaP vaccine. The acellular pertussis vaccine has a rate of adverse events similar to that of a TD vaccine (a tetanus–diphtheria vaccine containing no pertussis vaccine). The DTaP vaccine is given in several doses, with the first recommended at 2 months and the last recommended between the ages of 4 and 6 years of age.

Before the introduction of pertussis vaccines, an average of 178,171 cases of whooping cough were reported in the United States, with peaks reported every 2–5 years; more than 93% of reported cases occurred in children under 10 years of age. After vaccinations were introduced in the 1940s, incidence fell dramatically to less than 1000 by 1976. However, incidence rates have increased since 1980. In 2012, rates in the United States reached a high of 41,880 people— which was the highest it had been since 1955, when the numbers reached 62,786.

In the 1970s and 1980s, controversy erupted over whether the killed whole-cell pertussis vaccine caused permanent brain injury. However, extensive studies showed no connection of any type between the DPT vaccine and permanent brain injury. The alleged vaccine-induced brain damage proved to be due to an unrelated condition, infantile epilepsy. However, before the refutation, a few well-publicized anecdotal reports of permanent disability were blamed on the DPT vaccine and gave rise in the 1970s to anti-DPT movements. In the United States, low profit margins and an increase in vaccine-related lawsuits led many pharmaceutical companies to stop producing the DPT vaccine by the early 1980s. In 1982, the television documentary *DPT: Vaccine Roulette* depicted the lives of children whose severe disabilities were incorrectly blamed on the DPT vaccine. The ensuing negative publicity led to many lawsuits against vaccine manufacturers. By 1985, vaccine manufacturers had difficulty in obtaining liability insurance. The price of the DPT vaccine skyrocketed, leading providers to curtail purchases and limiting availability. Only one manufacturer remained in the United States by the end of 1985. To rectify the situation, Congress in 1986 passed the National Childhood Vaccine Injury Act (NCVIA), which established a federal no-fault system to compensate victims of injury caused by mandated vaccines. The majority of claims that have been filed through the NCVIA have been related to injuries allegedly caused by the whole-cell DPT, not the DaPT vaccine.

In 2010, 10 infants in California died from whooping cough, and health authorities declared there to be an epidemic, encompassing 9120 children. Demographic analysis identified a significant overlap in communities, with a cluster of nonmedical child exemptions and cases. The number of exemptions varied widely among communities. In some schools, more than three-fourths of parents filed for vaccination exemptions. Vaccine refusal based on nonmedical reasons, and personal belief exacerbated this outbreak of whooping cough, as it did with the Disneyland measles outbreak (see p. 98).

Although vaccination is the preferred method for prevention of whooping cough, antibiotics may be used to treat whooping cough in those who have been exposed and are at risk of severe disease. Antibiotics (erythromycin, azithromycin, or trimethoprim/sulfamethoxazole) are useful if started within 3 weeks of the initial symptoms, but otherwise, they have little effect in most people. In children less than 1 year old and women who are pregnant, antibiotics are recommended within 6 weeks of the onset of symptoms.

Chicken Pox

Chicken pox is not caused by a poxvirus (such as *Vaccinia*) but by a herpes virus *Varicella-Zoster*. Chicken pox was not recognized as a separate illness until the mid-1900s. Indeed, for many years, it was considered a milder form of smallpox. Before the development of the vaccine for chicken pox, the virus infected 4 million people each year in the United States and 100 million people worldwide.

The virus, *Varicella-Zoster*, is shed from the pustules in the skin of an infected individual and is spread from person to person by the airborne route. Infection first occurs in the tonsils and then in lymphocytes, and then, the virus moves to the skin. Over a period of 2 weeks, there is fever and malaise and finally the appearance of the rash that lasts about a week. Recovery usually results in immunity to reinfection. Although considered a mild disease, complications can be severe, with encephalitis, hepatitis, pneumonia, and bacterial skin infections due to *Staphylococcus* and *Streptococcus*. About 75% of those infected develop latent infections that persist for a lifetime, and when reactivated (usually in people over 50 years of age), the infection spreads from the site of latency (sensory ganglia) down the nerve to the skin, resulting in the painful and itching skin rash, shingles.

In 1951, Thomas Weller's 5-year-old son Peter came down with chicken pox. Weller broke open the blisters, collected the pus, and was able to grow the virus in human fibroblast cultures. Although he was the first to grow *Varicella* in tissue culture Weller found it difficult to serially propagate the virus and so was unable to produce a vaccine. In 1974, Michiahi Takahashi in the Microbiology Department at Osaka University isolated and attenuated the virus from a 3-year-old boy named Oka. Like Weller, he took fluid from the blister and passed it 11 times in human embryonic lung fibroblasts at low temperature, followed by 12 passages in guinea pig embryo cells and then 10 passages in WI-38 and MRC-5 human cells. Because there is no animal model for chicken pox, Takahashi and colleagues took a risky approach in administering the candidate vaccine to healthy children; luckily, it was found to be immunogenic and safe. It was licensed for use in Japan and Korea in 1988, with 1.8 million children vaccinated by 1993. Maurice Hilleman at Merck also tried to isolate and attenuate the virus but was unsuccessful, and so, he developed the Oka strain as a vaccine for the U.S. market in 1995. Today's live varicella vaccine is the Oka strain and is produced by Merck and GlaxoSmithKline. Before the chicken pox vaccine became available, chicken pox caused about 10,000 hospitalizations and 100 deaths annually in the United States. Since 1995, one dose of the vaccine has been recommended as part of the standard immunization schedule for children aged 1–12 years in the United States. Because of immunization, chicken pox has become much less common as an illness in children and with few fatalities. Indeed, by 1999, when 60% of toddlers received the vaccine, only 1 in 10 kids was getting chicken pox. For adults and children over 13 years of age, two routine doses are prescribed. The varicella vaccine was combined with measles, mumps, rubella (MMR) in 2006; however, with reports of febrile seizures in 2008, the combination measles, mumps, rubella, varicella (MMRV) was not recommended for use. The attenuated *Varicella* vaccine (Zostavax) also protects against shingles.

Parenthetically, it should be noted that infections with chicken pox can be treated with acyclovir.

Polio

During the first half of the twentieth century, no illness inspired more dread and panic than did polio. Summer was an especially bad time for children, when it was known as the "polio season." Children were among the most susceptible to paralytic poliomyelitis (also known as infantile paralysis). Many victims were left paralyzed for life. When exposed to the poliovirus in the first months of life, infants usually showed mild symptoms, because they were protected from paralysis by maternal antibodies still present in their bodies; however, as hygienic conditions improved and fewer newborns were exposed to the virus, paralytic poliomyelitis began to appear in older children and adults, who did not have any immunity. Perhaps, the most famous victim of polio was President Franklin Delano Roosevelt, who at the age of 39 contracted the disease, which left him crippled for life.

The poliovirus enters the body through the mouth and initially multiplies in the intestine. During the early stages of infection, the symptoms are fever, fatigue, headache, vomiting, stiffness in the neck, and pain in the limbs. Most infected patients recover; however, in a minority of patients, the virus attacks the nervous system. One in 200 infections leads to irreversible paralysis (usually in the legs). Among those paralyzed, 5%–10% die when their breathing muscles become immobilized. Although the names most associated with a vaccine for polio are Jonas Salk and Albert Sabin, their work would not have been possible without the studies of John Enders.

Enders, the son of a banker, was born on February 10, 1897, in West Hartford, Connecticut. He was educated at St. Paul's School in Concord, New Hampshire. After finishing school in 1915, he went to Yale University, but in 1917, he left his studies to become, in 1918, a pilot in the U.S. Air Force, with the rank of Ensign. After World War I, he returned to Yale and was awarded a BA degree in 1920. He then went into business in real estate in Hartford, but after becoming dissatisfied with this, he entered Harvard University. For 4 years, he studied English literature and Germanic and Celtic languages, with the idea of becoming a teacher of English. He did not decide on a life in microbiology, until he shared a room in a boarding house with several Harvard University medical students, one of whom was working with the legendary bacteriologist/immunologist Professor Hans Zinsser. Enders was captivated by the stories the medical students told of their laboratory work with Zinsser, and when Enders attended some of Zinsser's lectures and read his writings, he abandoned his graduate studies in literature. In 1929, a year before he earned his doctorate, Enders became Zinsser's teaching assistant. At that time, Zinsser was trying to develop a vaccine against the bacteria-like rickettsiae that cause typhus, Rocky Mountain spotted fever, and scrub typhus. Rickettsiae are similar to viruses in that they are smaller than bacteria and cannot be grown in culture like bacteria; instead, rickettsiae require a living cell for their growth and reproduction. In Zinsser's laboratory, Enders learned the methods of cultivating large quantities of the rickettsia in minced chick embryo tissues. After Enders spent 15 years with Zinsser in the Department of Bacteriology and Immunology at Harvard Medical School, studying mumps, and where he prepared a vaccine against Feline panleukopenia virus, he established his own laboratory at the Children's Hospital in Boston. There, two former medical school roommates, Weller and Fred Robbins, joined Enders. Before World War II, Enders and Weller were able to grow the vaccinia virus in chick embryo tissues in flask cultures or in roller tubes. In 1948, Weller succeeded in growing mumps virus in minced amniotic membranes obtained from aborted fetuses. With support from the National Foundation for Infantile Paralysis, Enders suggested that Weller and Robbins inoculate some roller tube cultures with poliovirus that was stored in the freezer. Weller set up tissue cultures containing fibroblasts from human foreskin and embryonic tissues from stillborn or premature babies who had died shortly before birth in the Boston Lying-In Hospital. Some of these were seeded with throat washings from his son Peter, suffering from chicken pox (*Varicella*), and one of Weller's particular interests. The remaining roller tubes were inoculated with a mouse adapted strain of type 2 polio. With Enders's encouragement, Weller maintained the cultures for long periods of time and changed the media weekly. The chicken pox virus failed to grow well; however, when the culture fluid from the polio culture was injected into the brains of mice, much to their surprise, they found that the mice became paralyzed, indicating that the poliovirus was multiplying in the roller tube cultures. This was a historic first, because until then, no one had been able to grow poliovirus in nonnervous human tissue, in part, because no one had tried and because the dogma was that poliomyelitis was exclusively a disease of the central nervous system and infected only neural tissues. They repeated their work with cultures of nervous tissue as well as skin, muscle, kidney, and intestine. These seminal observations were published in January 1949 in *Science* magazine: "Cultures of the Lansing strain of poliomyelitis virus in cultures of various human embryonic tissues." Later, they were able to grow poliovirus types 1 and 3 in similar culture. Because the polioviruses kill the cells in which they grow and multiply, and this was obvious microscopically, they had a convenient assay for virus multiplication and did not need to inoculate monkeys (for types 1 and 3) or mice (for type 2 poliovirus). In addition, it was possible by using this in vitro assay to assess the presence or absence of virus-specific antibodies obtained from monkeys that had been infected with poliovirus.

In 1954, the Nobel Prize for Physiology or Medicine was awarded for this work to Enders, Weller, and Robbins. As was characteristic of Enders, who could have easily been the sole recipient, the Nobel Prize was shared with his younger collaborators, who he said were full participants. Enders was a scientist with a green thumb and a great heart. He was unwavering in his honesty and was exceptional in his generosity, sharing reagents, cultures, viruses, and know-how with all who requested them. Within 4 years of the Enders, Weller, and Robbins publication, Jonas Salk was able to report success for a polio vaccine containing a formalin-killed preparation of the three types of poliovirus.

Jonas Salk was born in New York City in 1914. His parents were Russian-Jewish immigrants, who, despite lacking formal education themselves, were determined to see their children succeed, and encouraged them to study hard. When he was 13 years old, Salk entered Townshend Harris High School, a public school for intellectually gifted students that was "a launching pad for the talented sons of immigrant parents who lacked the money—and pedigree—to attend a top private school." In high school, "he was known as a perfectionist . . . who read everything he could lay his hands on," according to one of his fellow students. Students had to cram a 4-year curriculum in just 3 years. As a result, most dropped out or flunked out, despite the school's motto "study, study, study." Of the students who graduated, however, most would have the grades to enroll in the City College of New York (CCNY), noted for being a highly competitive college. At the age of 15, Salk entered CCNY, intending to study law, but at his mother's urging, he put aside aspirations of becoming a lawyer and instead concentrated on classes necessary for admission to medical school. Salk managed to squeeze through the Jewish admissions quota and entered the NYU Medical School in 1934. According to Salk: "My intention was to go to medical school, and then become a medical scientist. I did not intend to practice medicine, although in medical school and in my internship, I did all the things that were necessary to qualify me in that regard." During medical school, Salk was invited to spend a year in researching influenza. After completing medical school and his internship, Salk returned to the study of influenza viruses. He then joined his mentor, Dr. Thomas Francis, as a research fellow at the University of Michigan. There, at the behest of the U.S. Army, he worked to develop an influenza vaccine. By 1947, Salk decided to find an institution where he could direct his own laboratory. After three institutions turned him down, he received an offer from the University of Pittsburgh, with a promise that he would run his own laboratory. He accepted the offer, and in the fall of that year, he left Michigan and relocated to Pennsylvania. The promise, though, was not quite what he expected. After Salk arrived at Pittsburgh, "he discovered that he had been relegated to cramped, unequipped quarters in the basement of the old Municipal Hospital." Salk, a driven, obstinate, and self-assured individual, began to secure grants from the Mellon Family, and over time, he was able to build a working virology laboratory, where he continued his research on flu vaccines. It was in Pittsburgh that Salk began to put together the techniques that would lead to his polio vaccine. Salk used the Enders group's technique to grow poliovirus in monkey kidney cells. Then, he purified the virus and inactivated it with formaldehyde but kept it intact enough to trigger the necessary immune response. Salk's research caught the attention of Basil O'Connor, president of the National Foundation for Infantile Paralysis (now known as the March of Dimes Birth Defects Foundation) and then President Franklin D. Roosevelt's lawyer. Salk's killed injectable polio vaccine (IPV) was tested first in monkeys and then in patients who already had polio at the D.T. Watson Home for Crippled Children (now the Watson Institute). The tests were successful. Next, the vaccine was given to volunteers who had not had polio, including Salk, his laboratory staff, his wife, and their children. The volunteers developed antipolio antibodies and none had adverse reactions to the vaccine. Finally, in 1954, national testing began on 1 million children, aged 6–9 years, who were known as the Polio Pioneers: half received the IPV, and half received a placebo. One-third of the children, who lived in areas where vaccine was not available, were observed to evaluate the background level of polio in this age group. On April 12, 1955, the results were announced: the IPV was safe and effective.

After Salk made his successful vaccine, five pharmaceutical companies stepped forward to make it, and each of those companies was permitted to sell the vaccine to the public. One company, Cutter Laboratories, in Berkeley, California, made it badly. As a result of Cutter not filtering out the cells in which the poliovirus was growing, some virus escaped the killing effects of the formaldehyde treatment. As a consequence, more than 100 thousand children were inadvertently injected with live, dangerous polio. Worse still, those children injected with the Cutter-produced vaccine were able to infect 200 thousand people, resulting in 70,000 mild cases of polio, 200 people severely and permanently paralyzed cases, and 10 deaths. It was the first and only man-made polio epidemic and one of the worst biological disasters in American history. Federal regulators quickly identified the problem with Cutter's vaccine and established better standards for vaccine manufacture and safety. Cutter Laboratories never made another polio vaccine.

Within 2 years of the Salk vaccine's release, over 100 million Americans and millions more around the globe had been immunized. In the 2 years before the vaccine was widely available, the average number of polio cases in the United States was more than 45,000. By 1962, that number had dropped to 910. Salk neither patented the vaccine, nor did he earn any money from his discovery, preferring to see it distributed as widely as possible.

Following the announcement that Salk's IPV worked, Americans named hospitals after him; schools, streets, and babies were also named after Salk. Universities offered him honorary degrees, and countries issued proclamations in his honor. He was on the radio, on the cover of *Time* magazine, and made TV appearances. The vaccine's success made Salk an international hero, and during the late 1950s, he spent time refining the vaccine and establishing the scientific principles behind it. By 1960, however, Salk was ready to move on. Salk's dream was to create an independent research center, where a community of scholars interested in different aspects of biology—the study of life—could come together to follow their curiosity. In 1960, the dream was realized with the establishment of the Salk Institute for Biological Studies in La Jolla, California. At the institute, Salk tried to prepare killed vaccines against HIV and cancer, without success. He died of heart failure in 1995.

Albert Sabin developed the "other" polio vaccine. Sabin was born in Bialystok, Russia (now a part of Poland) on August 26, 1906. In 1921, Albert and his family immigrated to the United States to avoid the pogroms and the rabid anti-Semitism prevalent in Imperial Russia. Because the Sabin family was poor, Albert would not have been able to acquire a degree in higher education, if Albert's uncle, a dentist, had not agreed to finance his college education, provided that he study dentistry. In 1923, Sabin entered NYU as a predental student but switched to medicine after reading Sinclair Lewis's medical novel *Arrowsmith* and Paul deKruif's *Microbe Hunters*. Thus began a lifelong interest in virology and public health. In 1928, on graduation from NYU with a BS degree, he enrolled in the NYU School of Medicine and completed an MD in 1931—the year of a great polio epidemic in New York City and the worst since 1916. He then spent 2 years as an intern at Bellevue Hospital. He was a National Research Council Fellow at the Lister Institute in London, where studied virology, and in 1935, he was on the staff at the Rockefeller Institute, where he worked in the virus laboratory of Peter Olitsky. Shortly before Sabin's arrival at Rockefeller, Olitsky was warned that hiring him would be a mistake, since Sabin had a difficult personality. Olitsky was unmoved and regarded Sabin as a genius, and having worked successfully with geniuses before, he was anxious to have Sabin as an associate. At that time, Olitsky's laboratory was concerned with immunity to viruses, particularly poliomyelitis. At Rockefeller, Sabin worked tirelessly, refusing to take even Sundays or holidays off. He worked with infinite patience, had most careful technique, engaged in precise planning, and carried out detailed and elaborate recording of observations; he was especially incisive in the analysis of a problem and conducted skillful tests, with rigid controls. Sabin's interest in active immunization was nurtured by another scientist at Rockefeller, Max Theiler, whose work was focused on developing a live-virus vaccine for yellow fever. In 1938, the first field trials of Theiler's live-attenuated yellow fever vaccine were being tested in Brazil, and their success had a profound effect on Sabin. (Theiler was awarded the Nobel Prize in 1951 for the yellow fever vaccine.)

In 1939, Sabin established his own research laboratory at the University of Cincinnati Children's Hospital. World War II interrupted Sabin's work on polio vaccines. He served in the U.S. Army Medical Corps and developed experimental vaccines against dengue fever and Japanese encephalitis virus. At war's end, he returned to Cincinnati and to polio research.

Sabin had a longstanding interest in polio, and his first published article was in 1931 on the purification of the poliovirus. This was an important step in that it provided for the possibility of the oral route of infection and was contrary to the prevailing belief that the disease was primarily neurological. In 1936, Sabin (in collaboration with Olitsky) was able to grow the poliovirus in human embryo brain tissue.

One January day in 1948, Hilary Koprowski, then at the Lederle Laboratories in Pearl River, New York, macerated mouse brain material in an ordinary kitchen blender. He poured the result—thick, cold, gray, and greasy—into a beaker, lifted it to his lips, and drank. It tasted, he later said, like cod liver oil. He suffered no ill effects. Koprowski had set out to attenuate the poliovirus through adaptation to mouse brain. Starting with a type 2 mouse strain, he achieved attenuation of neurovirulence in monkeys. After ingesting the orally administered vaccine, he arranged to vaccinate 20 mentally disabled children in collaboration with the physician in charge of the institution in which they resided. The ethical justification was the fear of poliovirus entering the institution, a common occurrence at the time. Although this first trial showed safety and immunogenicity of the strain, the presentation of the results at a later scientific meeting was greeted with shock because of the audacity of the work. Two years later, Koprowski received a call from Letchworth Village, a home for mentally disabled children in Rockland County, New York. Fearing an outbreak of polio, the home asked him to vaccinate its children. In February 1950, in the first human trial of a live polio vaccine, Koprowski vaccinated 20 children there. At that time, approval from the federal government was required to market drugs but not to test them. Seventeen of the children developed antibodies against polio. (The other three turned out to have antibodies already.) None of the children experienced complications. Unfortunately, the father of a child receiving the vaccine developed paralysis and died. On autopsy, virus was recovered from the brain tissue and the vaccine was withdrawn. Although Koprowski was the first to produce a live oral polio vaccine (OPV) and his strains were tested extensively in the former Belgian Congo, his native Poland, and elsewhere, they were never approved for use in the United States because they were regarded as too virulent.

With the production of attenuated poliovirus strains, it became apparent to Sabin that oral exposure might be a promising direction for development of a vaccine. He was convinced that an oral vaccine would be more easily administered and better tolerated as a public health measure. Further, he felt that a live oral vaccine would mimic the natural infection to produce an asymptomatic infection in the gut that would stimulate the immune response and lead to the production of a systemic immune response. By 1954, Sabin had obtained three mutant strains of the poliovirus that appeared to stimulate antibody production without paralysis. Sabin entered into an agreement with Pfizer to produce the OPV, and the pharmaceutical company began to perfect its production techniques in its UK facilities. From 1957 to 1959, Sabin successfully tested the OPV on human subjects—himself, his family, research associates, and hundreds of prisoners from the nearby Chillicothe Penitentiary. Since during this time the Salk vaccine was being used in the United States, Sabin was unable to get support for a large-scale field trial for his OPV. However, because polio was widespread in the USSR, Sabin was able to convince the Soviet Union Health Ministry to conduct trials with his OPV. It was used in Russia, Estonia, Poland, Lithuania, Hungary, and East Germany, and by 1959, over 15 million Soviets, mainly children, had been given the Sabin live oral vaccine.

On the morning of April 24, 1960, more than 20,000 children in the greater Cincinnati area lined up to receive the Sabin OPV in its first public distribution in the United States. An additional 180,000 children in the surrounding area received the vaccine during the next several weeks in what became to be known as "Sabin Oral Sundays." In 1960, Sabin published a landmark article "Live, Orally Given Poliovirus Vaccine" in the *Journal of the American Medical Association*. In the article, he described the results of these large studies with children under the age of 11 years. Owing to the self-limiting nature of the poliovirus, a second dose was needed to achieve full protection. In 1961, the trivalent OPV (administered as drops or on a sugar cube) was licensed in the United States, and by 1963, it was the polio vaccine of choice. Between 1962 and 1964, more than 100 million Americans of all ages received the Sabin OPV. By the 1980s, after large public health field trials were conducted and found to be successful, the WHO adopted a goal of polio eradication. In 1994, the WHO declared that naturally occurring polio had been eradicated from the Western Hemisphere. By 1995, 80% of children worldwide had received the requisite three doses of the vaccine in the first year of life, and it is estimated that it prevented half a million cases of polio annually. During his lifetime, Sabin staunchly defended his OPV, refusing to believe that it could cause paralysis. Despite this belief, the risk, though slight, does exist. As a consequence, in 1999, a federal advisory panel recommended that the United States return to the Salk IPV because of its lower risk in causing disease. Based on a decade of additional evidence, this recommendation was affirmed in 2009.

Sabin was very severe and demanding of himself and of those with whom he worked. He monitored everything. He knew everything that happened in the laboratory during that day, and you would hear about it the next morning if things were not right. He had rigorous standards. Although he never saw his research as a race to the finish line, he once said that his mission was to kill the killed (Salk) vaccine, because he believed that a live oral vaccine was superior. Although there was rivalry between Salk and Sabin, in truth, Sabin simply did not have a very high opinion of Salk as a scientist. He was just another guy. Sabin was a stubborn but eloquent speaker, and it was often difficult to defeat him in scientific arguments. On the Salk vaccine, he once declared that it was "pure kitchen chemistry."

Sabin published more than 350 scientific papers. Although he never received the Nobel Prize, he received numerous other awards, including over 40 honorary degrees, the U.S. National Medal of Science, the Presidential Medal of Freedom, the Medal of Liberty, the Order of Friendship Among Peoples (Russia), the Lasker Clinical Research Award, and election to the U.S. National Academy of Sciences (1951). In 1970, he became the president of the Weizmann Institute. He retired from full-time work in 1986 at the age of 80, although he continued publishing until his death from congestive heart failure in 1993. He is buried in Arlington National Cemetery.

Of all the pioneering polio researchers, only Enders, Weller, and Robbins were awarded the Nobel Prize. Salk, Koprowski, and Sabin were never selected for the honor. The reasons for this remain controversial. In the case of Koprowski, it may be because his vaccine was not a universal success. Examination of the Nobel Archives reveals that Dr. Sven Gard, Professor of Virology at the Karolinska Institute, convinced the Nobel Committee to name Enders and his colleagues as recipients of the 1954 Prize, because as he wrote, "the discovery had had a revolutionary effect on the discipline of virology." Salk was nominated for the Prize in 1955 and 1956. For the first time, it was decided to wait for the results of the clinical trial of Salk's killed polio vaccine, which was in progress. In 1956, Gard wrote an eight-page analysis of Salk's work, in which he concluded, "Salk has not in the development of his methods introduced anything that is principally new, but only exploited discoveries made by others." He wrote: "Salk's publications on the poliomyelitis vaccine cannot be considered as Prize worthy." Few of the scientific societies honored Salk, and he received little recognition from his peers. Some in the scientific community considered his posing with movie stars and giving television interviews a behavior to be unbecoming for a prominent researcher. Although Salk was the public's darling, he remained a pariah in the scientific community. He received the Lasker Award but was never elected to the U.S. National Academy of Sciences, possibly because Salk was blackballed by Sabin, who sniped, "He never had an original idea in his life." Salk's rejection by the academy may also have been fueled by jealousy of his success in the public arena. In effect, Salk had broken two of the commandments of scientific research. Thou shalt give credit to others. Thou shalt not discuss one's work in newspapers and magazines.

The syphilitic man, 1496, by Albrecht Dürer. (Courtesy of the Wellcome Institute Library, London.)

Chapter 8

The Great Pox Syphilis and Salvarsan

It was the fall of 1932 and syphilis was rampant in small pockets of the American South. The U.S. Public Health Service began a study of the disease and enlisted 399 poor, black, share-croppers living in Macon County, Alabama, all with latent syphilis. Cooperation was obtained by offering financial incentives such as free burial service … on the condition that they agreed to an autopsy; the men were also given free physical exams, and a local county health nurse, Eunice Rivers, provided them with incidental medications such as "spring tonics" and aspirin whenever needed. The men (and their families) were not told that they had syphilis, instead they were told they had "bad blood" and annually a government doctor would take their blood pressure, listen to their hearts, obtain a blood sample, and advise them on their diet so that they could be helped with their "bad blood." However, these men were not told they would be deprived of treatment for their syphilis and they were never provided with enough information to make anything like an informed decision. The men enrolled in The Tuskegee Syphilis Study (as it was formally called) were denied access to treatment for syphilis even after 1947 when the antibiotic penicillin came into use (see p. 137). They were left to degenerate under the rav-ages of tertiary syphilis. By the time the study was made public, largely through James Jones's book *Bad Blood: The Tuskegee Syphilis Experiment* and the play "Miss Evers' Boys," 28 men had died of syphilis, 100 others were dead of related complications, at least 40 wives had been infected, and 19 children had contracted the disease at birth.

The Tuskegee Study was designed to document the natural history of syphilis, but for some it came to symbolize racism in medicine, ethical misconduct in medical research, and paternal-ism by physicians and government abuse of society's most vulnerable—the poor and unedu-cated. On May 16, 1997, the surviving participants in the Study were invited to a White House ceremony and President Bill Clinton said: "The United States government did something that was wrong—deeply, profoundly, morally wrong. It was an outrage to our commitment to integ-rity and equality for all our citizens. Today all we can do is apologize but you have the power. Only you have the power to forgive. Your presence here shows … you have not withheld the power to forgive. I hope today and tomorrow every American will remember your lesson and live by it."

A Look Back

In 1996, the 500th anniversary of the arrival of syphilis in England was "celebrated." Syphilis—"the Great Pox" as the English called it—was a disease that from 1493 onward swept over Europe and the rest of the world, including China, India, and Japan. The claim was made that this new disease was brought to Naples by the Spanish troops sent to support Alphonso II of Naples against the French king, Charles VIII. Charles VIII launched an invasion of Italy in 1494, and besieged the city of Naples in 1495. During the siege his troops, consisting of 30,000 mercenaries from Germany, Switzerland, England, Hungary, Poland, and Spain as well as those from France, fell ill with the Great Pox and this forced their withdrawal. It is generally believed that with the disbanding and dispersal of the soldiers of Charles VIII, who themselves had been infected by the Neapolitan women, the pox spread rapidly through Europe. In the spring of 1496 some of these mercenaries joined Perkin Warbeck in Scotland and with the support of James IV invaded England. The pox was evident in the invading troops. Within 5 years of its arrival in Europe the disease was epidemic: it was in Hungary and Russia by 1497; and in Africa and the Middle East a year later. The Portuguese carried it around the Cape of Good Hope with the voyages of Vasco de Gama to India in 1498. By 1505 it was in China, in Australia by 1515, and in Japan by 1569. European sailors carried the Great Pox to every continent save for Antarctica. Syphilis was so ubiquitous by the nineteenth century that it was considered to be the AIDS epidemic of that era.

But if the story of the outbreak of this pox in the army of Charles VIII is accurate, how did his men contract this disease? There are two main theories called the Columbian and the pre-Columbian (or by some, the anti-Columbian). Let's consider the Columbian theory first. Christopher Columbus (1451–1506) visited the Americas and on October 12, 1492, arrived in San Salvador. He set sail for home 3 months later on January 16, 1493, and arrived in Spain in March of 1493, carrying with him several natives of the West Indies. The crew of 44, upon arrival in Spain, disbanded and some of these joined the army of Charles VIII. The first mention of the disease occurs in an edict by Emperor Maximilian of the Holy Roman Empire at the Diet of Worms in 1495, where it is referred to as "the evil pox." Twenty-five years later in a book published in Venice and authored by Francisco Lopes de Villalobos, it was claimed that syphilis had been imported into Europe from the Americas. Favoring this idea was the severity of the outbreak—indicative of a new import. Indeed, from 1494 to 1516, the first signs were described as genital ulcers, followed by a rash, and then the disease spread throughout the body affecting the gums, palate, uvula, jaw, tonsils, and eventually destroying these organs. The victims suffered pains in the muscles and there was early death—an acute disease. From 1516 to 1526 two new symptoms were added to the list: bone inflammation and hard pustules. Between 1526 and 1560, the severity of symptoms diminished and thereafter its lethal effects continued to decline, but from 1560 to 1610, there was another sign: ringing in the ears. By the 1600s, "the Great Pox" was an extremely dangerous infection, but those who were afflicted did not suffer the acute attacks that had been seen in the 1500s.

In the 1700s, syphilis was a dangerous but not an explosive infection. By the end of the 1800s, both its virulence and the number of cases declined. Even so, the numbers were by no means trivial: by the end of the nineteenth century it was estimated that 10% of the population of Europe was infected, and by the early twentieth century mental institutions noted that one-third of all patients could trace their neurological symptoms to syphilis.

Clearly, either the people were developing an increased resistance or the disease's pathogenicity was changing.

What produced this dramatic outbreak of syphilis? Some have suggested that it was a new disease introduced into a naïve population and that the increased rate of transmission by sexual means transformed what once had been a milder disease into a highly virulent one (such as happened with HIV). Another hypothesis was that European syphilis was derived from yaws and initially infection was by direct contact; one simple means might have been by mouth-to-mouth kissing as this was the more common practice of greeting (rather than a handshake) in Tudor

England. In addition to kissing, it was suggested, infection could have been passed by shared drinking cups. In this case, the infectious canker (chancre) occurred on the lips and tongue, and went unnoticed or was mistaken for more benign diseases such as cold sores or impetigo. Perhaps another reason for the rapid spread of the pox may have been that precautions against transmission were not observed since the later stages would not have been recognized as being associated with the earlier and more infectious stages, and even these stages—chancre and rash—might have been considered nothing more than minor, self-healing skin disorders.

Treatments for "the Great Pox" varied with the times. George Sommariva of Verona, Italy, tried mercury for the treatment of "the French pox." And by 1497 mercury was applied topically to the suppurating sores or it was taken in the form of a drink. The treatment came to be called "salivation" because the near poisoning with mercury salts tended to produce copious amounts of saliva. Another treatment was guaiacum (holy wood) resin from trees (*Gauiacum officianale* and *G. sanctum*) indigenous to the West Indies and South America. In actual fact, the resin was a useless remedy, but it was popularized probably to lend further credence to the American (Columbian) origins of the pox. However, few adherents of the Columbian theory paid attention to the fact that guaiacum was introduced as a treatment in 1508, a good 10 years before the first mention of the West Indian origin of syphilis. Further, the claim for the Columbian origin of syphilis would seem to be weakened by several written reports that the crew of Columbus and the Amerindians were healthy. There are others, however, who suggested (but never showed) that some of Columbus's sailors did have syphilitic lesions. Perhaps the best evidence for the Columbian origin of syphilis lies in the bones and teeth: bone lesions—scrimshaw patterns and saber thickenings on the lower limbs of adults and notched teeth in children—are diagnostic of syphilis and have been found in the skeletal remains of Amerindians; until recently, no such characteristic bone lesions were found in skeletons in Europe or China that were dated to be older than the 1500s, or in Egyptian mummies. But in 2000, there was a report of 245 skeletons that were unearthed from a medieval monastery, known as Blackfriars, in Hull, England. Carbon dating of the bones established the date of death to be between 1300 and 1420—at least 70 years before Columbus's voyage. Some paleoanthropologists believe these Blackfriar skeletons show the telltale signs of syphilis whereas others dispute the evidence and contend that the bone lesions are more like those that result from a related (but nonvenereal) disease called yaws. But, this refutation seems not to hold up in the face of the finding of notched teeth in the remains of individuals found in the port city of Metaponto, Italy, and dated to 600 BC as well as in those who died in Pompeii during the eruption of Mount Vesuvius (79 AD). Some believe this evidence is proof that syphilis was present in Europe for thousands of years.

Port cities with their characteristics of high sexual activity and prostitution provide the locale for the possibility of high transmission rates of sexually transmitted diseases. It has been hypothesized that when syphilis became a disease of cities there were changes in social habits, the wearing of clothing, and an absence of sharing of eating utensils and improved hygiene, the propagation of the milder form was reduced, and this allowed transmission of only the more virulent form. According to this view, syphilis would have existed in Europe and Asia in a milder form prior to 1493, but then it became virulent. Exactly when and how venereal syphilis arose continues to be debated. The solution to the question of the origin of "the Great Pox" may come only when the bones of Columbus and members of his crew are found and can be shown to contain, by DNA analysis, the genetic signature of the germ of syphilis.

Spirochete Discovered

What is the causative agent of the disease syphilis? Early observers believed syphilis was God's punishment for human sexual excesses. Public bathhouses were closed and there was distrust between friends and lovers. Others believed that syphilis had an astrologic basis. In 1484 Mars, Jupiter, and Saturn were in conjunction with Scorpio, the constellation most commonly associated with sexuality. But as early as 1530, Girolamo Fracastoro (1483–1553) recognized that this disease was contagious, and he called the disease "syphilis" after a fictitious shepherd, named Syphilis, who got it by cursing the gods. Fracastoro described the earliest stages of syphilis as small ulcers on the genitals followed by a skin rash; the pustules ulcerated and the person suffered with a severe cough that eroded the palate. Sometimes the lips and eyes were eaten away, or ulcerated swellings appeared, and there were pains in the joints and muscles. Fracastoro theorized that syphilis was a result of "seeds of contagion" but for 400 years no one saw the "seeds" that caused the symptoms of the great pox. Then in 1905, Fritz Schaudinn (1871–1906) and Erich Hoffmann (1869–1959) in Germany identified a slender, corkscrew-shaped bacterium, a spirochete ("spiro" meaning "coiled"; "chete" meaning "hair"), in the syphilitic chancres. When Hideyo Noguchi isolated the same bacterium from the brains of patients suffering with insanity and paresis, it became clear that all of the stages of the disease were linked to one kind of spirochete. Because of its shape Schaudinn and Hoffmann called the microbe *Treponema* ("trep" meaning "corkscrew" and "nema" meaning "thread" in Latin) and because it stained so poorly they named its kind *pallidum* (from "pallid" meaning "pale"). *T. pallidum*, has no spores, cannot be cultured on bacteriologic media, but it can be grown in experimental animals such as rabbits and guinea pigs. However, human beings are the only natural host for *T. pallidum*. It divides slowly (~24 h) and is quite fragile requiring a moist environment. Other spirochetes, related to *Treponema*, cause relapsing fever and Lyme disease (*Borrelia burgdorferi*).

The Disease Syphilis

Since there is enormous variation in the disease symptoms, syphilis has been called "The Great Imitator." How do we know about the clinical course of a syphilitic infection? In 1890–1910, 1404 untreated Swedish patients were studied—the Oslo Study. Seventy-nine percent of those with late-stage syphilis developed neurosyphilis, 16% had destructive ulcers (gummas), 35% had cardiovascular disease, and in 15% of men and 8% of women syphilis was the primary cause of death. Between 1917 and 1941, the Rosahn study of cadavers of those who died from syphilis found that late-stage syphilis was evident in 39%, neurological complications were found in 9%, and cardiovascular disease in 83%. Between 1932 and 1972, the now infamous

Tuskegee Syphilis Study was designed to track untreated syphilis in poor, rural, and uneducated black males in Macon County, Alabama, where the overall infection rate was 36%. The study involved 399 men with latent syphilis compared with a control group of 201 healthy men. Signs of cardiovascular disease were found in 46% of the syphilitics, 24% had an inflammation of the aorta whereas the controls had 24% and 5%, respectively. Bone disease was found in 13% of the syphilitics but only in 5% of the controls. The greatest differences were seen in the central nervous system; in the syphilitics 8% had signs of disease compared to 2% in the controls. After 12 years, the death rate in the syphilitics was 25% versus 14%, at 20 years it was 39% versus 26%, and at 30 years 59% versus 45%.

The chancre stage is the earliest clinical sign of disease. After initial infection, that is ~21 days (range 3–90 days) after initial contact, a painless, pea-sized ulcer appears at the site of spirochete inoculation—a chancre. The chancre is a local tissue reaction and can occur on the lips, fingers, or genitals. If untreated, the chancre usually disappears within 4–8 weeks leaving a small inconspicuous scar. The individual frequently does not notice either the chancre or the scar, although there may be lymphadenopathy. (The lesion is larger than that of smallpox hence the disease has been named "the Great Pox.") At this stage, kissing or touching a person with active lesions on the lips, genitalia, breasts, and through breast milk can spread the infection.

The secondary or disseminated stage usually develops 2–12 weeks (mean 6 weeks) after the chancre; however, this stage may be delayed for more than a year. *T. pallidum* is present in all the tissues but especially the blood, and there is a high level of syphilis antigen. Serologic tests (such as the Wasserman, Venereal Disease Research Laboratory [VDRL] test, and rapid plasma reagin [RPR]) are positive. There is now a general tissue reaction—headache, sore throat, a mild fever, and in 90% of the cases a skin rash. The skin rash may be mistaken for measles or smallpox or some other skin disease. The highly infectious secondary stage does not last very long. Then the patient enters the early latent stage, in which he or she appears to be symptom-free, that is, there are no clinical signs. Indeed the most dangerous time is during the early latent stage for the infected individual can transmit to others but appears to be disease-free. Rarely, transmission occurs by blood transfusion, because of low incidence and the fact that the spirochetes do not survive longer than 24–48 h under blood bank storage conditions.

The infection continues to progress and after about 2 years the late latent or tertiary stage develops. In tertiary syphilis there are still spirochetes present in the body but the individual is no longer infectious. For all intents and purposes, individuals are not infectious through sexual contact 4 years after initial contact. Tertiary syphilis develops in one-third of untreated individuals 10–25 years after initial infection. The disease then becomes chronic. In about 20% of the cases destructive ulcers (gummas) appear in the skin, muscles, liver, lungs, and eyes. In 10% of the cases the heart is damaged and the aorta is inflamed. In severe cases, the aorta may rupture causing death. In 40% of untreated cases, the spinal cord and brain become involved causing incomplete paralysis, complete paralysis and/or insanity, accompanied by headaches, pains in the joints, impotence, and epileptic seizures. The remaining cases have asymptomatic neurosyphilis. Most untreated patients die within 5 years after showing the first signs of paralysis and insanity.

Syphilis can be transmitted from the mother to the developing fetus via the placental blood supply resulting in congenital syphilis; this is most likely to occur when the mother is in an active stage of infection. If the mother is treated during the first 4 months of pregnancy the fetus will not become infected. Fetal death or miscarriage usually does not occur until after the 4th month of pregnancy at the earliest. Repeated miscarriages after the 4th month are strongly suggestive of but not unequivocal proof for syphilis. A diseased surviving child may go through the same symptoms as the mother or there may be deformities, deafness, and blindness. Of particular significance is that some offspring who are congenitally infected may have Hutchinson's triad: deafness, impaired vision, and a telltale groove across peg-shaped teeth, first described in 1861 by the London physician Jonathan Hutchinson.

Diagnosing Syphilis

In the chancre stage, syphilis is rarely diagnosed by growth and isolation of *T. pallidum* itself. Instead, examination of fluid from the lesion by dark field microscopy and clinical findings are the basis for disease determination. The Wassermann test is a serologic reaction that originally used extracts of tissues infected with *T. pallidum*, however, when it was found that uninfected tissues reacted positively with syphilitic sera these antibodies (called reaginic) were considered nonspecific. The nonspecific reaction appears to be due to the external waxy layer of the spirochete. Despite this these serologic tests, which are cheap, easy to use, and quick, continue to play a role in screening patients especially those in high-risk populations as well as in evaluating adequacy of treatment. The most commonly used tests are the VDRL and the RPR card tests. Both become negative 1 year after successful treatment for primary syphilis and 2 years after successful treatment of secondary syphilis. Patients that are infected with both HIV and syphilis commonly have a serologic response with very high titers. Fluorescent antibody absorbed tests and hemagglutination assays are used to confirm a positive nontreponemal, that is, a nonspecific test.

Salvarsan

By the end of the nineteenth century, it was estimated that 10% of the population of Europe had syphilis and by the early twentieth century it was estimated that one-third of all patients in mental institutions could trace their neurological symptoms to the disease. The roll call of famous syphilitics includes Gustave Flaubert, Charles Baudelaire, Guy de Maupassant, Eduoard Manet, Henri de Toulouse Lautrec, and Paul Gaugin. The art dealer Theo van Gogh was a long term sufferer, and his brother Vincent may have been driven to his famous ear-lopping incident by syphilis. It has been speculated (and without much foundation) that Ludwig van Beethoven, Franz Schubert, Friedrich Nietzche, Fyodor Dostoyevsky, Napoleon Bonaparte, Edgar Allen Poe, Niccolò Paganini, Abraham Lincoln, Oscar Wilde, and Lord Randolph Churchill (Winston's father), were syphilitics. The syphilitic club is joined by Colette; Al Capone; Adolph Hitler; and Karen Blixen, a.k.a. Isak Dinasen, author of *Out of Africa*, who caught it from her philandering husband.

In 1883, Paul Ehrlich was working with Emil von Behring in Robert Koch's laboratory in Berlin investigating the relationship of neutralizing antibodies to toxins and was able to standardize diphtheria antitoxin so it could be used in the treatment of human infections. As a result of this work the German government established the Royal Institute for Experimental Therapy in Frankfurt to provide such antitoxins and vaccines and appointed Ehrlich as director in 1899. Visits to the Hoechst factory near Frankfurt brought Ehrlich face to face with German dye works where a profusion of synthetic analgesics, antipyretics, and anesthetics were being made. It seemed logical to him that since such substances were effective by acting on specific tissues (much in the way dyes did) that it should be possible to synthesize other small molecules that would act differentially on tissues and parasites. He recognized that immunotherapy—the use of vaccines—was a matter of strengthening the defense mechanisms of the body, whereas with drug therapy there would be a direct attack on the parasite. And, he also observed that in immunotherapy it required an animal to make the large, unstable protein molecule (antibody) whereas in drug therapy a small and stable molecule could be made in the laboratory.

Ehrlich wrote that "curative substances—a priori—must directly destroy the microbes provoking the disease; not by an 'action from distance,' but only when the chemical compound is fixed by the parasites. The parasites can only be killed if the chemical has a specific affinity for them and binds to them. This is a very difficult task because it is necessary to find chemical compounds, which have a strong destructive effect upon the parasites, but which do not at all, or only to a minimum extent, attack or damage the organs of the body. There must be a planned chemical synthesis: proceeding from a chemical substance with a recognizable activity, making derivatives from it, and then trying each one of these to discover the degree of its activity and effectiveness. This we call chemotherapy." Ehrlich's first chemotherapeutic experiments were carried out in 1904 with mice infected with the trypanosomes that cause African sleeping sickness. He was able to cure these mice of their infection by injections of a red dye he called trypan red. This was the first man-made chemotherapeutic agent. At first, trypan red aroused some interest, however, because the drug was inactive in human sleeping sickness Ehrlich turned his attention to organic arsenicals. He began with the compound atoxyl—an arsenic-containing compound that was supposedly a curative for sleeping sickness, but he found the drug was useless: it destroyed the optic nerve so that when patients were treated with the drug they not only were not cured of sleeping sickness they became blind.

Most of Ehrlich's contemporaries ridiculed him calling him "Dr. Phantasmus"; furthermore, they were unimpressed with his dye-based research and thought nothing of value would come from it. Indeed, for 5 years, Ehrlich was unable to produce a single drug that was of use in humans. But in 1905 all that changed. That year, Schaudinn and Hoffmann described the germ of syphilis, and shortly thereafter Sahachiro Hata working at the Kitasato Institute for Infectious Disease in Tokyo discovered how to reproduce the disease in rabbits. It was intuition rather than logic that led Ehrlich to believe that arsenicals would kill *Treponema*. (His reasoning was that since trypanosomes and treponemes are both active swimmers they must have a very high rate of metabolism and an arsenical would kill the parasite by crippling their energy-generating ability.) And now, thanks to Hata's rabbit model for human syphilis Ehrlich's drugs could be tested for their chemotherapeutic effectiveness. In the spring of 1909, Professor Kitasato of Tokyo sent his pupil, Dr. Hata, to Germany to work with Ehrlich and a year later Hata successfully treated syphilis in rabbits with a dioxy-diamino-arseno-benzene compound named 606. The synthesis of 606 was reported in 1912, patented, manufactured by Hoechst, and sold by them under the name "salvarsan." The newspapers took up the announcement of a cure for syphilis in humans and overnight Ehrlich became a world celebrity.

Salvarsan was a yellow, acidic powder, packaged in 0.6 g ampoules. One difficulty with using it was the problem in dissolving this powder and neutralizing it under aseptic conditions prior to injection in order to avoid oxidation that could alter its toxicity. Ehrlich issued precise directions for its preparation in sterile water, neutralization, and advised that intravenous injection must be carried out without delay. These directions were often not adhered to and the resulting deaths attracted much unfavorable publicity. Further, it was dangerous to make salvarsan because the ether vapors used in its preparation could cause fires and explosion. There were problems with side effects of salvarsan—such as gastritis, headaches, chills, fever, nausea, vomiting, and skin eruptions—and, at times, the syphilis had progressed so far that it was not effective as a cure. Moreover, salvarsan was not selective when given as a single dose and so treatment had to be spread out over several months. This meant that fewer than 25% of the patients ever completed the course of treatment. Despite these problems, salvarsan and its successor (neoarsphenamine or 914, a more soluble derivative of 606) remained the best available treatment for 40 years. Although Ehrlich was awarded the 1908 Nobel Prize (with von Behring) for work on immunity, he regarded his greatest contribution to be the development of a "magic bullet," salvarsan. At the Nobel ceremony, he modestly said, "My dear colleagues for seven years of misfortune I had one moment of good luck!"

The green mold *Penicillium notatum*, the source of penicillin. (Courtesy of Wellcome Images.)

Prontosil,
Pyrimethamine,
and Penicillin

In the early morning hours of October 26, 1939, Gerhard Domagk was roused from slumber by the incessant ringing of the telephone. Picking up the phone, he heard the caller in a Swedish-accented voice say: "You have been awarded the Nobel Prize for Physiology or Medicine for the discovery of the antibacterial effects of Prontosil." Domagk was stunned and elated. However, his euphoria was soon tempered by the fact that Fuhrer Adolph Hitler had forbidden any German to accept a Nobel Prize or deliver the lecture in Stockholm, Sweden. Domagk was forced to sign a letter addressed to the appropriate Nobel committee that stated: It was against his nation's law to accept, and the award also represented an attempt to provoke and disobey his Fuhrer. At the end of World War II, Domagk was reinvited to travel to Stockholm to accept on behalf of the Karolinska Institute and the King the Nobel Prize consisting of a gold medal and diploma and to deliver his speech. The money, 140,000 Swedish crowns—a rich award worth several years of salary for most scientists—he would never see, since in accordance with the rules of the Nobel Foundation unclaimed prize money reverts back to the main prize fund. In the introductory speech, delivered by Professor N. Svartz of the Karolinska Institute, in 1947, she said: "During the past 15 to 20 years a great deal of work has been carried out by various drug manufacturers with a view to producing less toxic but at the same time therapeutically effective ... preparations. Professor Gerhard Domagk ... planned and directed the investigations involving experimental animals. The discovery of Prontosil opened up undreamed of prospects for the treatment of infectious diseases. What Paul Ehrlich dreamed of, and also made a reality using Salvarsan in an exceptional case has now, through your work become a widely recognized fact. We can now justifiably believe that in the future infectious diseases will be eradicated by means of chemical compounds."

Prontosil

As with Paul Ehrlich's discovery of "magic bullets" such as methylene blue and Salvarsan, the development of Prontosil began with the synthesis of a group of red-colored azo dyes. Initially, the azo dyes were of interest to the chemists at IG Farben for their ability to contribute to fastness for dyeing wool not for their antimicrobial activity! Although azo dyes similar to Prontosil had been synthesized almost a decade earlier (1919) by Heidelberger and Jacobs at the Rockefeller Institute, they were discarded after they showed poor antimicrobial activity in the test tube. Indeed, it was fortuitous that in 1933–1935, Domagk in trying to understand the lack of correlation between test tube and animal antibacterial tests, resorted to examining drugs such as Prontosil in an animal model.

In 1921, Domagk who was 26, graduated from medical school. His first position was at the hospital in Kiel where it was possible to observe, first hand, infectious diseases especially those that were caused by bacteria. However, the hospital was not well equipped and there was poor access to microscopes; the equipment for chemistry was limited and laboratory facilities were nil. At the time, physicians knew what bacteria were and that they might cause disease; however, there was little understanding as to how physicians actually produced the disease state. Moreover, they had no clue as to how to stop the disease-producing bacteria save for carbolic acid antisepsis.

Unfortunately, carbolic acid (phenol) was indiscriminate in action, killing not only the bacteria but normal tissues as well. The harsh antiseptics could only be used for external use and once an infection had been established, little could be done to control it. What was needed were substances for use within the body after an infection had established itself, a kind of internal antisepsis. Eager to learn the latest techniques and findings, Domagk moved to the Institute of Pathology at the University of Greifswald in 1923, and he then moved to what appeared to be a better post at the University of Munster in 1925. Two years later, he found himself dissatisfied with his ability to carry out research in a university setting. Then fortune smiled on him.

In 1927, Domagk met with Carl Duisberg who told him Bayer was expanding its drug research and was looking for someone with experience in both medicine and animal testing to screen their chemicals for medical effects in animals. Domagk was, according to Duisberg, the perfect candidate and not only would he be given a large laboratory but he would be appointed as director of the Institute of Experimental Pathology. At Bayer, Domagk's laboratories were carved out of the space occupied by the head of the Tropical Disease unit, Wilhelm Roehl. Roehl, a young physician with an intellectual's high forehead but with a gangster's heavy features—picture Edward G. Robinson in a white laboratory coat—was a former assistant to Paul Ehrlich at Speyer House. He had joined Bayer at the end of 1909 (see p. 13). Roehl showed Domagk how he ran the animal tests and how his results with the compounds he had screened were correlated. Indeed, Roehl already had some success in developing Plasmochin (=Plasmoquin) using his malaria-infected canary screening method and the compounds that had been provided to him by the Bayer chemists. Things were beginning to look up for Domagk who was to take over the work initiated by the chemist Robert Schnitzer at Hoechst (already a part of the IG Farben cartel) in his search for a drug against a generalized bacterial infection, particularly the hemolytic streptococcus (*S. pyogenes* that caused meningitis, rheumatic fever, middle ear infections [otitis media], tonsillitis, scarlet fever, and nephritis). During World War I, this streptococcus was also responsible for the many deaths after wounding, and oftentimes hemolytic streptococcal infections were a consequence of burns and scalding. In addition, in concert with the influenza virus it caused pneumonia responsible for many deaths during the worldwide epidemic in 1918–1919. The appearance of streptococci in the blood of a patient, septicemia, was often a prelude to death. Indeed, in 1929, while Roehl was traveling in Egypt he noticed a boil on his neck while shaving. It turned out to be infected with *S. pyogenes*. Roehl developed septicemia for which there was no cure and in a few days Roehl, age 48, was dead.

In the 1920s, Rebecca Lancefield at the Rockefeller Institute in New York, discovered that there were many types of hemolytic streptococci; however, not every one of them was a danger to human health. Based on this Domagk decided to select a particularly virulent strain of hemolytic streptococcus, one isolated from a patient who had died from septicemia. After inoculation into mice, this strain reliably killed 100% within 4 days. Domagk reasoned that only an exceptional drug would enable the infected mice to survive his "drop-dead" test. Over time, Domagk would test thousands of compounds synthesized by the chemists Joseph Klarer (hired by Bayer at the same time as Domagk) and Fritz Mietsch (who had come to the company to work with Roehl and had synthesized Atabrine; see p. 15). In their syntheses,

Klarer and Mietsch followed the trail blazed by Ehrlich and they concentrated their efforts on dyes that specifically bound to and killed bacteria. Gerhard Domagk examined several of the acridines synthesized by Klarer and Mietsch for activity against streptococcal infections in mice. Optimum activity was found with an acridine with an amino group on the carbon opposite the nitrogen atom rather than to the adjacent ring as was the case with pamaquin. This 9-aminoacridine when given orally or by injection to streptococcal-infected mice was able to control an acute infection; however, this particular nitroacridine was not potent enough to be used clinically. However, another analog was found to be more effective as an antibacterial; it was marketed as Entozon (or Nitroakridin 3582) and found use in the treatment of bovine mastitis and uterine infections.

Domagk tested two other classes of compounds that had been reported clinically to have antibacterial properties: gold compounds and azo dyes. In 1926, Hoechst's gold compound, gold sodium thiosulfate (Sanochrysine), had been shown to be effective in bovine tuberculosis and it protected mice against strep. Domagk also found it to be effective in the "drop-dead" test; however, it could not be used for treating humans because when the dosage necessary to cure patients of their strep infections was used, it resulted in kidney damage. Domagk then tried an azo dye because the synthesis of azo dyes was simple, chemical variations were relatively easy to make, and they were less toxic. In addition, they were easier to make than acridines, which had already been proved negative in Domagk's rigorous test system. Domagk tried the azo dye, "Pyridium," and although it inhibited bacterial growth it turned the urine color to red and this would be unacceptable to patients. In September 1931, an azo compound made by Klarer, called chrysodine, which incorporated a chlorine atom cured mice even after oral administration, but it was not powerful enough for clinical trials. A year later, Klarer began attaching sulfur atoms and he produced *para*-amino-benzene-sulfonamide.

This compound, number 695 in the Klarer azo series, completely protected mice when given by injection as well as by mouth. It protected mice at every dose level; however, much to Domagk's surprise, it did not kill the streptococci in the test tube, but only in the living animal did it work. Furthermore, it was specific acting only on streptococci and not on other bacteria. Klarer kept making other modifications. Domagk found that one, number 730, a dark-red azo dye, almost insoluble in water, was tolerated by mice at very high doses and it was without side effects. It was like all the others in the series of azo dyes in that it was selective for streptococci and did not kill tuberculosis-causing bacteria, or the pneumonia-causing pneumococi or staphylococci. On December 27, 1932, Domagk concluded that this compound, named Streptozon, was the best for treating streptococcal infections. And as was the practice at IG Farben, a patent was filed under the names of the chemists, Klarer and Mietsch, not the pathologist Domagk. (And as was also the practice, only the chemists received royalties.) Streptozon was shown to be clinically effective; however, there was a drawback: the brick-red powder could be made into tablets and ingested, but was impossible to dissolve. A liquid formulation was needed, and in 1934, Mietsch produced a beautiful port-red fluid that could be injected. It was just as active as the powder. Streptozon solubile as it was called, was patented, and renamed Prontosil because it worked quickly.

Prontosil was given the chemical name sulfamidochrysodine, although this was almost never used. Prontosil treatment saved many lives from the death sentence imposed by *S. pyogenes* including Domagk's own 6-year-old daughter, Hildegarde. The little girl had fallen down the stairs at home and the needle she was carrying punctured and broke in her hand. The needle was removed surgically; however, the wound became infected and she developed septicemia. To avoid amputation of the arm and possible death Domagk gave his daughter Prontosil by mouth and rectally. She recovered. Apart from the 1935 introduction of Prontosil by IG Farben, and the thousands of sulfonamides synthesized at Elberfeld during the 5 years after the first recognition of the their antibacterial properties, no other major developments in this field came from Germany. The initiative passed to the British, the French, the Americans, and the Swiss.

In April 1935, when the Pasteur Institute in Paris, France, requested a sample of Prontosil, and this was denied, Ernest Forneau (1872–1949), a leading French chemist and head of the Laboratory of Therapeutic Chemistry at the institute, instructed his staff to decipher what they could of the German patent application and then synthesize it. (Under French law, Prontosil could only be patented as a dye and not a medicine.) Using a method described in the literature in 1908 by Paul Gelmo of Vienna, who had prepared it for his doctoral thesis, the French found a way to duplicate Prontosil's synthesis and began to manufacture it. It was marketed as Rubriazol, because of its red color. There remained, however, a puzzle. Why didn't Prontosil follow Paul Ehrlich's axiom: *corpora non agunt nisi fixata* (a drug will not work unless it is bound) that guided much of chemotherapy? Domagk wrote: "It is remarkable that in vitro it shows no noticeable effect against *Streptococcus* or *Staphylococcus*. It exerts a true chemotherapeutic effect only in the living animal."

Forneau's group at the Pasteur Institute continued to study the drug and found something unexpected: mice infected with streptococci but treated with an azo-less compound, that is, pure sulfanilamide survived. Simple sulfanilamide was colorless and unpatentable; it was, however, as effective as the red wonder drugs, Prontosil or Rubriazol. The French finding explained why attaching sulfa to many types of the azo dyes resulted in activity against strep, whereas dyes without sulfa were less active. The mystery of Prontosil's activity in the live animal and not in the test tube was solved. In order to become active Prontosil had to be split to release the 4-amino-benzene-sulfonamide, and this was due to enzymes in the body (Figure 9.1). In the

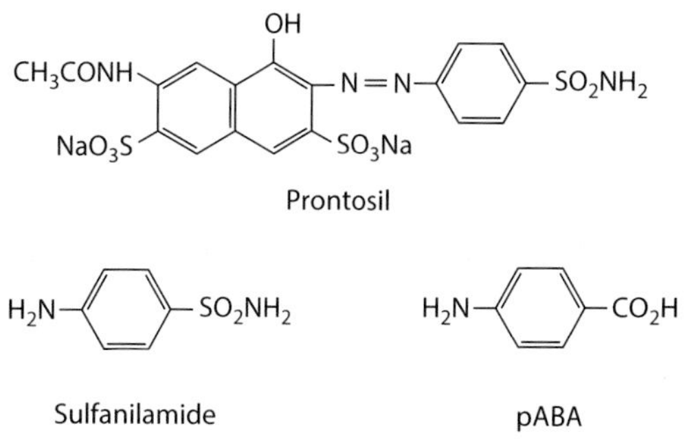

FIGURE 9.1 The relationship of prontosil with sulfanilamide and that of pABA.

test tube where there were no enzymes to split the Prontosil, no sulfa could be released. Thus, Prontosil had to be "bioactivated." In short, the azo dye stained whereas the sulfa moiety cured.

Drug firms were able to produce chemical variants of sulfa that could work much better, were less toxic, and had a wider range of activity against many different kinds of bacteria. May and Baker in England produced a version with an attached aminopyridine, called M & B 693 or sulfapyridine. Sulfapyridine was also shown to be effective against streptococcal, meningococcal, staphylococcal, and gonococcal infections. It had unprecedented efficacy against mice with lobar pneumonia, and in human trials it reduced the mortality from 1 in 4 to 1 in 25. Among others, it saved the life of Great Britain's Prime Minister Winston Churchill. On December 11, 1943, as Churchill was flying to Dwight Eisenhower's villa in Tunis after an exhausting trip where he had conferred with Roosevelt and Stalin at Yalta to firm up plans for the invasion of Italy, he fell ill. At the age of 69, he was overworked, overweight, and mentally exhausted. By the time he arrived in Tunis, his throat was sore and his temperature was 101°F. An X-ray revealed a shadow on his lung, and as his lungs became more congested it suggested lobar pneumonia. He suffered two bouts of atrial fibrillation and an episode of cardiac failure. As he hovered near death, Churchill was given M & B 693, and it worked. By Christmas, he was back in action planning the invasion of France and within two weeks was back home. Most were convinced he owed his recovery to the new medicine, although he joked that in using M & B, he was referring to his physicians, Moran and Bedford. Churchill said, "There is no doubt that pneumonia is a very different illness from what it was before this marvelous drug was discovered." He might have also said it was sulfa's finest hour!

Prontosil proved that Ehrlich's "magic bullets" were possible. Sulfa encouraged the discovery of newer drugs, established research methods needed to find them, framed the legal structure under which they could be sold, and created a business model for their development. Today thousands of sulfa drugs have appeared in the market, some of them short-lived, but a number have stood the test of time. The number of patents issued for sulfa drugs was 1300 in 1940, and increased to 6500 in 1951. In 1943 in the United States alone, 4500 tons of sulfonamide were produced and millions of patients treated. Ironically, by the time Domagk received the Nobel Prize many drug makers seemed to be less interested in sulfa drugs, and they began to turn their attention to the more powerful antibiotics such as penicillin and streptomycin. Sadly, Domagk died in the spring of 1964 from a bacterial infection that did not respond to either sulfa drugs or antibiotics. The microbes that killed him were able to repel the chemically designed "magic bullets."

What remained unanswered by Domagk and Forneau was why sulfa drugs worked. In the 1940s, Donald Woods and Paul Fildes, working in London, found that sulfa was less a magic bullet than a clever imposter. They started from the observation that sulfa never worked as well in the presence of pus or dead tissue. Woods went looking for the mysterious antisulfa substance that interfered with its action and found it as well in yeast extract. He characterized the antisulfa substance as a small molecule, approximately the size of sulfanilamide. In fact, the mystery substance looked like sulfa's molecular twin, it was *para*-aminobenzoic acid (pABA) (Figure 9.1).

pABA is a chemical involved in the nutrition of certain kinds of bacteria, although today it is more familiarly associated as a sunscreen ingredient. Some bacteria can make pABA; however, others (such as streptococci) cannot. For those microbes unable to synthesize pABA from scratch it is an essential nutrient—a vitamin—and if their environment does not provide it, they starve to death. What Woods and Fildes showed was that sulfa worked because it was a molecular mimic of pABA, and when the sulfa was around the bacteria would try to metabolize it instead of pABA. However, the sulfa could not be utilized, and once bound to a specific bacterial enzyme (dihydropteroate synthase, DHPS) that enzyme could no longer function. In sum, sulfa was for those bacteria that required pABA an antibacterial antimetabolite.

Almost at the same time, that sulfanilamide was identified as the active principle of Prontosil, L. T. Coggeshall found it to be an effective antimalarial in monkeys infected with *P. knowlesi*. In 1941, Maier and Riley following the lead provided by Woods and Fildes showed that the activity of sulfanilamide against the chicken malaria *P. gallinaceum* (isolated by Emile Brumpt in 1935) could be reversed by pABA. Shortly thereafter, it was shown that sulfadiazine was active against *P. falciparum* and *P. vivax* in human subjects. And, the group of biochemists at Harvard University working under research contracts from the Committee on Medical Research, Board for the Coordination of Malaria Studies, showed that not only could pABA reverse the action of sulfadiazine but it was a requirement for the test tube growth of the human malaria *P. knowlesi*. Later studies demonstrated that depletion of host pABA (milk diet) inhibited malaria infections in monkeys and supplementation of the diet with pABA resulted in a resurgence of the infection.

In 1952, K. Eli Marshall, Professor of Pharmacology at Johns Hopkins, spoke factually about his work on sulfonamides, but not the excitement, triumph, and significance of the years 1941–1946 when he worked in chemotherapy and was a prime mover in the U.S. government's World War II malaria program:

> It was amusing to be a free lance … waiting for some accidental observation to point out a promising lead. Then in 1936 I began to read about Streptozon and how it cured human cases of streptococcal and staphylococcal septicemia. In 1919–1920 I had become interested in the chemotherapy of bacterial infections. Nothing came of this interest except an unpublished address before the St. Louis Section of the American Chemical Society. When successful bacterial chemotherapy arrived I was ready for it. As I see it, our significant contributions to bacterial chemotherapy were as follows. A simple, accurate, and specific method was devised for the determination of sulphonamides in blood and tissues. These had the effect of devising a rational basis of dosage—an initial loading dose and then a maintenance dose every 4 h day and night. Soon dosage of sulfonamides was based on blood concentrations rather than on the number of grams administered by mouth. As a result of the studies … new sulfonamide drugs were introduced into clinical use. The chemical and pharmacological properties of the sodium salt of sulfapyridine were first described from my laboratory. This compound was introduced clinically for intravenous use and was the precursor of the use of sodium salts of other sulphonamides.

Marshall might have also mentioned that various sulfonamides were very effective in suppressing the bird malaria *P. lophurae*. Discovered in a Borneo Fireback pheasant housed in the New York Zoo, its blood was used to infect white Pekin ducklings. Marshall found that optimal therapy depends upon the maintenance of a more or less constant blood concentration of these drugs, and that the complete inactivity of sulfonamides against *P. relictum* malaria in the canary may be due not only to a species difference in parasite susceptibility but also to the use of single daily doses of these drugs instead of maintenance of continuous blood concentrations. Indeed, later it was shown that several sulfonamide drugs, reported to be inactive in bird malaria when tested by a single dose schedule were found to be active when examined by the drug–diet method. There were further complications: the Roehl test involved administering small amounts of the test compound and the size of the canary precluded taking many samples of blood. Recently, it has been possible to show that malaria parasites can actually synthesize pABA from "scratch," that is, de novo. Further, all of the genes encoding the enzymes of the folate pathway are present in *P. falciparum* and the rodent malaria *P. yoelii*; yet, the parasites appear to be unable to synthesize sufficient quantities to survive *in vivo* and hence an extra source is needed.

How do sulfonamides work as antimalarial drugs? Sulfonamides are considered as "antifolates" (or antifols) because they inhibit the essential synthetic pathway to form the molecule, tetrahydrofolate (THF), which in turn acts as a methyl group carrier for the thymidylate synthase reaction critical to DNA synthesis (Figure 9.2). Specifically, DHPS, which catalyzes the

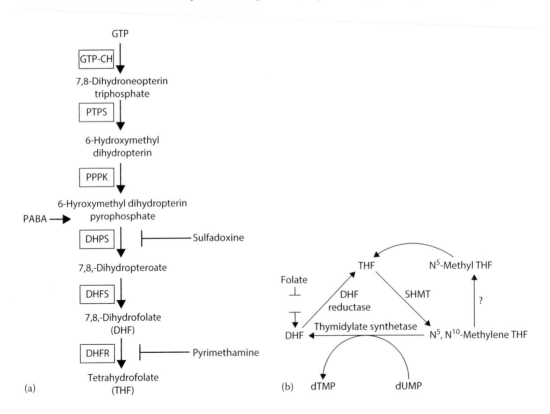

FIGURE 9.2 (a) The folate pathway and (b) the thymidylate cycle.

condensation of pABA with 6-hydroxymethyl dihydropterin pyrophosphate, is the site of action of sulphonamides (Figure 9.2a). Since humans and other backboned animals cannot synthesize folate from scratch, it must be obtained from the diet (hence it is a vitamin); as a consequence a block in THF synthesis by sulfonamide does not occur in humans. In contrast, spineless creatures, plants, and microbes including malaria parasites and *Toxoplasma* do synthesize folates using simpler molecules such as guanosine triphosphate (GTP), pABA, and the amino acid glutamate using a stepwise sequence of enzymes and when there is an interruption THF cannot be made. When this happens the synthesis of DNA cannot occur because the nucleotide thymidylate (dTMP) cannot be made.

A somewhat simple model for this, analogous to the molecular changes effected by a sequential action of enzymes to produce THF, is to think of it as occurring in a flowing river where there are sluice gates at various positions along the course of the river and each sluice gate is controlled by a gatekeeper. The gatekeepers control the flow along the length of the river. Opening and closing by a gatekeeper permits a regulated flow, but if a gatekeeper falls asleep at the switch then sluice gate cannot function properly. If it remains closed (blocked), flow downstream diminishes or ceases with an accumulation of water upstream of that particular gate, and flow beyond the gate is reduced. Sulfonamides, mimic pABA (Figure 9.1), and act by binding to a particular gatekeeper (an enzyme in this model), the enzyme DHPS so that the product dihydropteroate cannot be formed (Figure 9.2a). (In effect, the gatekeeper is "asleep" and this blocks downstream flow of the pathway.) Or, the mimic may be converted by DHPS to form a complex with pterin that is unusable for the next gatekeeper enzyme, dihydrofolate synthase (DHFS), and as a result there is depletion of the downstream pool of DHF. Further, because there is a greater affinity of sulfonamides for the parasite's DHPS, sulfonamide is both a specific and an effective antimalarial.

Pyrimethamine

George Hitchings (1908–1998) of the Burroughs Wellcome Company received the Nobel Prize in 1988, "for … discoveries of important principles for drug treatment." Hitchings theorized that it should be possible to alter the way cells grow and reproduce by substituting slightly different compounds for those occurring naturally. He thought it possible to fool the cell into thinking it was replicating itself when in fact it had been handed what Hitchings liked to call a "rubber doughnut." He believed that since all cells synthesize DNA and RNA it might be possible to use analogs of the nucleic acid bases that make up the DNA and RNA to stop cellular multiplication.

Hitchings's interest in nucleic acids stemmed from his PhD work at Harvard and the development of microanalytical methods for the purine bases to follow the metabolism of ATP. After completion of his PhD in 1933, during the depths of the depression, he experienced a 9-year period of impermanence both financial and intellectual with short appointments at Harvard's C. P. Huntington Laboratories in cancer research, the Harvard School of Public Health in nutrition research, and at Western Reserve University in electrolyte research.

Following this, Hitchings, now 39 years of age, joined the U.S. laboratories of Burroughs Wellcome, a subsidiary of the British company.

Sir Henry Wellcome founded the Wellcome Bureau of Scientific Research in London in 1913, to carry out research in tropical medicine. The job of Trust Director was offered to Andrew Balfour, who had been in charge of the Wellcome Tropical Research Laboratory in Khartoum since its creation in 1902. Although funding was derived from the commercial activities of the Wellcome Foundation Ltd. (later Wellcome PLC), including Burroughs Wellcome Inc., its wholly owned American subsidiary, the scientific staff were relatively free to establish their own lines of investigation. Henry Wellcome's tenet for his research laboratories had always been "Freedom of research—Liberty to publish," and this attracted some of the most talented scientists of the day to work there. After the bureau moved to its Euston Road site in London in 1934, it became known as the Wellcome Laboratories of Tropical Medicine. In 1965, these laboratories moved to the Beckenham site to be merged with the existing Wellcome Research Laboratories, which had been there since 1922. Wellcome also had a laboratory housed in a converted rubber factory in Tuckahoe, New York. It was there, in 1942, that Hitchings joined the Burroughs Wellcome Company as the sole member of the Biochemistry Department and he began to explore synthetic mimics (analogs) of pyrimidines and purines as inhibitors of DNA biosynthesis. At Burroughs Wellcome, as in many pharmaceutical company laboratories, the research goal was to design inhibitors for a particular biological target. The underlying principle was that even if there was not an absolute difference in the metabolism of parasite and the host, there were probably enough differences at the active or binding site of iso-functional enzymes (isozymes) to allow for differential inhibition. Hitchings hypothesized (based in part on the antimetabolite principle expressed by Woods and Fildes in 1940) that it should be possible to alter the way cells grow and develop by substituting analogs for the naturally occurring molecules.

At the time Hitchings began his work, none of the enzymes or the steps in the formation of nucleic acids were known and the deciphering of the double helix structure of DNA by Watson and Crick was a decade away. Hitchings began studying how harmless bacteria incorporated synthetic analogs of pyrimidines and purines. "The reasons for choosing pyrimidine derivatives as a basis for molecular modification turned out, however, to be largely fallacious. Resemblance was sought with earlier antimalarial compounds … based on the rather far-fetched analogy of sulfonamides as derivatives of aniline, with a side chain performing the same function as the basic side chain of older antimalarials. Pyrimidines were (also) looked upon favorably because they showed the characteristic resonance of quinacrine … and because there was a chance that they could interfere with … pyrimidine metabolism."

Hitchings, working in concert with Elvira Falco, then an assistant in the company's Bacteriology Department, and an organic chemist, Barbara Roth, used a folate-dependent strain of the bacterium *Lactobacillus casei*. Initially, more than 100 pyrimidine analogs were examined for their ability to inhibit the growth of *L. casei*. The most active was an analog of thymine, a 2,4-diaminopyrimidine. However, other work with *Streptococcus faecalis*, which showed that folinic acid (THF) was about 500 times more potent in reversing the inhibition than was folic acid (dihydrofolate, DHF), suggested that the analog was interfering somewhere in the biosynthesis of THF from DHF.

At Wellcome Laboratories in the United Kingdom, Leonard Goodwin (1915–2008), who had joined the company in 1939, was given the task of evaluating the compounds Hitchings group had prepared. Goodwin was born in north London, but soon after World War I, his family moved to Hampstead, where he spent a happy childhood. On the advice of his uncle, a pharmacist, he went to University College London (UCL) and qualified as a BPharm in 1935. He was then offered a post as a UCL demonstrator with a salary of £165. After 4 years, during which time he earned a BSc and MB, he looked for a job with better pay. He wrote to C. M. Wenyon, the director of Research at the Wellcome Laboratories of Tropical Medicine in Central London, asking if there are any job opportunities. A few days after the interview with Wenyon, Goodwin was called back and told: "I think I can find you something to do." The "something" was to measure the activity of a series of new antimony compounds which seemed to have promise for the treatment of kala-azar, a virulent tropical disease affecting vast areas of India then a part of Britain's empire. At the time, toxicology testing involved the calculation of a "chemotherapeutic index" based on the relationship between the maximum dose that can be tolerated by the body without causing death and the minimum dose that cures the infection. Goodwin felt the existing test to be too imprecise and he began measuring the effect of the antimony compounds on the livers of infected Syrian hamsters. Unlike European hamsters, the Syrian hamsters were highly susceptible to infection and within a year British soldiers were being treated with the antimony compound Pentostam. With Pentostam Len's career was truly launched. Subsequent to this, and with limited supplies of quinine during World War II, Wellcome began to direct its research to investigating drugs for the control and prevention of malaria. Goodwin set up a screen using *P. gallinaceum* infective for chickens to test promising compounds to replace quinine. Of the 300 substituted 2,4-diaminopyrimidines sent by Hitchings, Roth, and Falco, the most active ones were in the series substituted with a phenyl in the five position and the highest antimalarial activity was found with one compound optimized by Falco and another member of the Hitchings group, an organic chemist, Peter B. Russell who noted its resemblance to the antimalarial cycloguanil. It was 2,4-diamino-5-*p*-chlorophenyl-6-ethylpyrimidine (later named pyrimethamine and marketed as Daraprim).

When all the animal toxicology of pyrimethamine had been completed, Goodwin and another willing volunteer would take increasing doses and monitor the results. For example, Len took a dose that rose to 100 mg a day but this had to be stopped because it adversely affected the production of red blood cells in his bone marrow. As a result, the recommended and effective dose was established at 25 mg, once a week. On another occasion, he gave himself large doses of pyrimethamine for a year and while on a trip to Kenya with no other protection from malaria, he allowed *P. falciparum*-infected mosquitoes to feed on his arm. The drug was protective and he did not come down with malaria. During clinical trials, with other willing volunteers, it was found that a single large dose of pyrimethamine in human subjects produced peak serum concentrations in the first 2 h and declined thereafter. The antifolic acid activity of the serum from a human subject treated with pyrimethamine was confirmed using *L. casei*, and subsequently, it was found that the duration of antimalarial activity follows the antifolic acid activity of the blood.

A relative of malaria is *Toxoplasma gondii*. Domestic cats and their relatives are the only known definitive hosts of *T. gondii*. Large numbers of unsporulated oocysts are shed in the cat's feces. Oocysts take 1–5 days to sporulate in the environment and become infective. Intermediate hosts (birds and rodents) became infected after ingesting soil, water, or plant material contaminated with oocysts. Shortly after ingestion, the oocysts transform into rapidly multiplying tachyzoites. These tachyzoites localize in neural and muscle tissue and develop into tissue cyst stages, bradyzoites. Cats become infected after consuming birds or rodents harboring tissue cysts. (Cats may also become infected directly by ingestion of sporulated oocysts, although this is less effective than eating bradyzoite-infected meat.) Humans can become infected by any of several routes: eating undercooked meat of animals harboring tissue cysts, consuming food or water contaminated with cat feces or by contaminated environmental samples (such as fecal-contaminated soil or changing the litter box of an infected pet cat), blood transfusion or organ transplantation. (Humans are however accidental hosts.) In the human, the parasites form tissue cysts, most commonly in skeletal muscle, myocardium, brain, and eyes; these cysts may remain throughout the life of the host. Although 60 million Americans may harbor *T. gondii*, the vast majority will never experience symptoms. However, when acquired by pregnant women and with transmission to the fetus it is the cause of a tragic yet preventable disease in the offspring. In addition to the unfortunate outcome for infants and children are the emotional and economic burdens faced by the parents and society. It has been estimated that 500–5000 infants each year are born with congenital toxoplasmosis in the United States. Although the majority of infants appear to be healthy at birth, significant long-term sequelae may become obvious only months or years later. In the early 1950s, Daraprim was tested for its efficacy in treating *T. gondii* infections in experimental animals and in humans. In 1953, the U.S. Food and Drug Administration approved the drug.

In September 2015, Daraprim once again became big news when the *Los Angeles Times* reported: "Until last month, patients suffering from the parasite-borne disease toxoplasmosis which in its worst manifestations can cause blindness, neurological problems or death—had reasonable access to a remedy: a six-week, two-pill-a-day course of the drug Daraprim, at a cost of about $1,130." Then, an entrepreneurial company called Turing Pharmaceuticals acquired the exclusive rights to Daraprim and raised its price from $13.50 per pill to $750, bringing the total treatment cost to $63,000. For longer-term patients such as sufferers from HIV, the annual cost can go as high as $634,000, according to a joint statement from the Infectious Diseases Society of America and the HIV Medicine Association. With public outcry, Turing agreed to reduce the price.

How does Daraprim work to treat toxoplasmosis? *T. gondii* manufactures folates in a similar fashion to *Plasmodium* (see Figure 9.2a) and is mediated by six enzymes: GTP-CH (GTP cyclohydrolase), PTPS (pyruvoyltetrahydropterin synthase), PPPK (hydroxymethyl dihydropterin pyrophosphokinase), DHPS, DHFS, and DHFR (dihydrofolate reductase).

DHFR is the key enzyme that catalyzes the transformation of DHF to THF, a form of folate able to be utilized in the process leading to thymidine formation prior to its incorporation into DNA. The presence of thymidylate synthase (TS) as well as DHFR in *P. lophurae*, led my colleague Edward Platzer to suggest there was a thymidylate synthesis cycle (Figure 9.2b). In this proposed cycle, inhibition of DHFR by antifolates such as pyrimethamine (Daraprim) blocks the regeneration of THF resulting in an interruption in the synthesis of thymidylate, thereby stopping parasite DNA synthesis. It is particularly important to note that the basis of the chemotherapeutic effectiveness of pyrimethamine in malaria and *Toxoplasma* is the extremely tight binding of drug to the parasite DHFR, in contrast to that to the enzyme from the host. (It has been shown that it is point mutations in the parasite's DHFR gene that result in resistance by *Plasmodium* and *Toxoplasma,* thus leading to treatment failures.)

Penicillin

It was March 14, 1942. Anne Miller, a young 33-year-old mother, was dying from a streptococcal infection. For weeks, she had a fever that ranged from 103°F to 104°F. She could not eat and was losing weight; she went in and of coma. Blood transfusions and sulfa drugs all failed to produce a cure. In desperation, her physician contacted John Fulton (at Yale University), who in turn knew the chairman of the Committee on Chemotherapy in Washington, DC, which at the time controlled all the important medicines during World War II (see p. 16). After the call, the chairman authorized the pharmaceutical company Merck to release 5.5 g—a teaspoonful—of penicillin and half the entire amount available in the United States. Not knowing the proper dosage it was given in several small doses during the day and within 24 h the deadly streptococcal infection had been cleared from her blood. She convalesced for a month, received more Merck-manufactured penicillin, went home, lived a full and productive life, and in 1999, at age 90, died. Anne Miller was the first patient in the United States pulled from death's door by the "miracle drug" penicillin. Where did penicillin come from and who discovered it?

The discovery of penicillin is usually attributed to Alexander Fleming who was born in Scotland in 1881. Fleming was an unimposing short man with a shock of white hair, and a flattened nose, but with bright blue eyes. In 1901, he won a tuition scholarship to St. Mary's Hospital medical school, qualified as a physician in 1908, and a year after joined the Inoculation Department at St. Mary's where he remained for the next 49 years. In 1909, Fleming described a method for testing syphilis using only a drop or two of blood and a year after Ehrlich developed salvarsan (see p. 121). Fleming was giving shots of salvarsan to his private patients.

During World War I, as a Lieutenant in the British Army, Fleming experienced and was concerned with the manner in which open wounds were treated using cloth bandages soaked in the antiseptic carbolic acid. Indeed, of the 10 million soldiers killed in World War I, half died not from mustard gas or bullets or bombs but from bacterial infections. Fleming recognized that although the antiseptic was effective in killing the bacteria in the pus-laden surface wounds, the bacteria remained deeper within the wound and could cause tissue destruction and death. Surely, he reasoned, there must be some treatment that could prevent an infection from becoming established rather than overcoming it once it was already present. The search for a material to prevent infection now became the focus of Fleming's life's work.

In November 1921, Fleming was at St. Mary's and suffering with a head cold; he decided to put the mucus from his runny nose on to a Petri dish that had been seeded with bacteria. Later, when Fleming noticed that the colonies of bacteria were dissolving he tried to isolate the bacteriolytic substance which he believed to be an enzyme. Between 1922 and 1932, Fleming published eight papers on the substance he called lysozyme (and he found it also occurred in egg white, turnips, human tears, and bronchial mucus). Fleming had limited skills as a writer and his papers on lysozyme were short and had few details. Further, he was a dreadful speaker. Indeed, when he presented his work on lysozyme to the Medical Research Club those in the audience responded with silence. Most in attendance felt his discovery to be irrelevant especially so when he noted that lysozyme had no effect on the bacteria that caused serious illness. As a consequence, the papers Fleming published on lysozyme were soon forgotten.

In 1928, when Fleming was playing with microbes he made a serendipitous discovery. At the end of July and before he left on vacation he piled two or three dozen Petri dishes seeded with *Staphyloccocus* at the end of his workbench. Sometime after he returned in August or September he looked at the dishes his research assistant had not discarded and found that a blue-green mold had contaminated one of the dishes. But the odd thing was that within the sea of staphylococci there were no bacteria surrounding the mold. He mused, "Something in the contaminating mold made those bacteria lyse." It reminded him of lysozyme and was, as the baseball player Yogi Berra would have said, "It was deja vu all over again." Two months later when Fleming recorded his observation in his laboratory notebook and was in need of a photograph to prove the original observation he grew the mold for 5 days at room temperature and then he seeded the dish with staphylococci; only then did he put the Petri dish in the incubator. In this case there was an obvious zone of inhibition of bacterial growth. Why did Fleming make such a change in the sequence of events? Because when he tried to reproduce the original finding by putting the mold directly on the lawn of bacteria and keeping the dish at room temperature there was no inhibition. If he did the reverse, that is, if he grew the staphylococci first and then added the mold there was no zone of inhibition and nothing noteworthy to photograph. (Later, Fleming and other researchers would find that staphylococci growth would be inhibited only if the mold had been grown first; this is because the mold grows best at room temperature whereas for the best growth of the bacteria body temperature was needed.)

Fleming's story of stray spore of a green mold landing on a dish of staphylococci and producing an inhibitory substance, which he called penicillin, after the mold's name *Penicillium*, was not published until 1944, and it appears to be more fiction than fact. In 1968, Ronald Hare who worked in the Inoculation Department at St. Mary's during the same time as did Fleming dismissed the notion of a stray spore coming in through the window of Fleming's 10 foot by 12 foot laboratory since the window was seldom if ever opened and no good bacteriologist would ever work near an open window for fear of contaminating the cultures. A far more likely source, according to Hare, was that the mold spores came up through the stairwell from the floor below where an Irish scientist, C. J. LaTouche, was working with molds. Further, Hare found that LaTouche had in his collection the very same *Penicillium* mold that contaminated Fleming's staphylococci plate. Continued attempts by Fleming and others to repeat his original find failed repeatedly. Hare suspects that this is because from July 28 to August 7 (the time Fleming was away), the temperature in London rose above 68°F only twice. Because *Penicillium* grows best at room temperature but the bacteria need body temperature (98°F) for good growth if Fleming had not incubated the dish with the staphylococci they would not have grown but the dish with the mold would have had a chance to establish itself before the weather turned warmer, and only then would the bacteria grow. Is Hare's speculation correct? We shall never know since Fleming's paper describing the action of the mold on the staphylococci is short on several important details such as the species of staphylococci, how long the dishes had been on the workbench, the medium of the culture, and whether and when he incubated the Petri dish. Some suggest that Fleming was actually looking for lysozyme activity (not penicillin) when he cultured the mold. When he tested the mold against bacteria known to be sensitive to lysozyme as well as *Staphylococcus* and other pathogenic bacteria and found that the mold worked on pathogens on which lysozyme had no effect; Fleming had found something totally unexpected—a mold that produced an antibacterial substance.

Fleming was fond of saying, "When you have acquired knowledge and experience it is very pleasant … to be able to find something nobody had thought of." Fleming continued to work with *Pencillium*. He grew the mold, made large batches of broth, and collected the "mold juice"; to his disappointment he found the antibacterial properties of the juice were quickly lost meaning penicillin was unstable. In addition to *Staphylococcus* he found that penicillin was effective against *Streptococcus*, as well as the bacteria that cause gonorrhea, meningitis, and diphtheria. However, penicillin was not effective against the tubercle bacillus and was nontoxic when injected into healthy rabbits. But, for some strange reason Fleming never tested the mold juice on an animal infected with *Staphylococcus* or *Streptococcus*.

On February 13, 1929, Fleming delivered a paper, "A Medium for the Isolation of Pfeiffer's Bacillus" (presumably about penicillin inhibiting some kinds of bacteria) to the Medical Research Club. The evening was a replay of his 1921 talk on lysozyme in that there was little reaction by the audience. In May, he submitted a paper "On the antibacterial action of cultures of penicillin with special reference to their use in the isolation of *B. influenzae*" to the *British Journal of Experimental Pathology* and it was published a month later.

It is likely that Fleming would have remained an obscure bacteriologist and penicillin's main application would be in the isolation of penicillin-insensitive bacteria in mixed cultures instead of a useful therapeutic were it not for the subsequent work at the Dunn School of Pathology at Oxford University by Howard Florey, Ernst Chain, and Norman Heatley.

Howard Florey was born in Adelaide, Australia, on September 24, 1898. He entered Adelaide Medical School in 1916, graduated first in his class in 1921, and then with a Rhodes scholarship left Australia for Oxford University (where he shared classes with another Rhodes scholar, John Fulton (see p. 137). At Oxford, he was taken under the wing of Professor of Physiology Charles Sherrington (who would in 1932 share the Nobel Prize in Physiology or Medicine for his work on the nervous system). Thanks to Sherrington's efforts, Florey was awarded a prestigious studentship at Cambridge University where he became interested in mucus secretions. After one year at Cambridge he traveled to the United States with a Rockefeller Foundation fellowship where he worked in several different laboratories, returned to Cambridge, and became lecturer in special pathology teaching physiology and doing research in pathology. Florey was short, with an arresting presence; a lean face, piercing eyes, and hair neatly parted in the middle. In 1935, he was appointed Professor of Pathology at the Dunn School of Pathology at Oxford University, where he was determined to revitalize its teaching and research. Florey was outspoken and had the roughness of a no-nonsense Australian. Within a week at Oxford, Florey recruited the German-born Ernst Chain, a Jewish immigrant with a 1930 PhD in biochemistry. Chain has been described as a better-kempt version of Albert Einstein with brushed back curly black hair and a moustache. He was flamboyant in appearance and in temperament. Chain studied for a second PhD at Cambridge University with biochemist Sir Frederick G. Hopkins, a 1929 Nobel laureate, working on snake venoms and the manner by which they cause central nervous system paralysis. Also joining Chain was Norman Heatley with a doctorate in biochemistry; his thesis was "The Application of Microchemical Methods to Biological Problems." Chain and Heatley were opposites in temperament, personality, and scientific approach. Heatley wrote meticulous notes whereas Chain took few notes. Heatley was reluctant to seek credit whereas Chain was quick to claim his due. Yet, despite these differences the two began to work together on cancerous tumors; they found no differences in the metabolism between normal skin and malignant cells. Even before they completed the studies on cancerous cells, Florey, with an abiding interest in mucus, asked Chain to study lysozyme because of his previous work on snake venom enzymes. Chain, in collaboration with an American Rhodes scholar Leslie Epstein, showed that lysozyme worked by digesting complex carbohydrates, polysaccharides. Florey recognizing Chain's interest in the action of lysozyme told him that since 1929 bacterial antagonism was an interest of his. Stimulated by Florey's interest, Chain's approach to the problem of bacterial antagonism was to read all the literature on the subject. Chain collected 200 references on growth inhibitors caused by the action of bacteria and molds; however, according to the literature nothing was known about the chemical nature of the inhibitory substances. Then one day in 1938, by sheer luck, Chain came across Fleming's paper on penicillin. Florey was already familiar with Fleming's work on penicillin since he had been the *British Journal of Experimental Pathology* editor of Fleming's paper. Chain thought that perhaps penicillin was also an enzyme and was deserving of further study. Chain tells the story that he took some *Penicillium* mold back to his lab and did some preliminary experiments and then brought this to the attention of Florey. Florey disputes this and claims the aim of the research was to study bacterial antagonism and that they started with penicillin because he already knew it was potent against staphylococci.

With the lysozyme study coming to an end and 3 days after Britain's entry into the war on September 6, 1939, Florey asked for monetary support from the MRC (Medical Research Council) and the Rockefeller Foundation for studies of antibacterial substances, specifically penicillin, suggesting there might be something of therapeutic value for humans. The MRC provided £25 and the possibility of £100 later and the Rockefeller granted $5000 (=£1200) per annum for 5 years. Now, the penicillin project became a team effort.

Heatley's first task was to find a more productive means for growing the *Penicillium* and then to devise a quick way to assay for the antibacterial activity of penicillin. The assay settled on was to make little wells in the agar in a Petri dish and then to seed the dish with bacteria. Whatever fluid to be tested was put into the wells and then the dish was incubated so the bacteria could grow. The distance the material diffused out of the well to inhibit the growth of the bacteria would be a measure of potency. Heatley then devised a better medium for the growth of the mold and was able to get greater production of penicillin. Chain found penicillin to be most stable from pH 5 to pH 8—close to neutrality. By mid-March of 1940, Heatley through prodigious effort was able to provide Chain with 100 mg of penicillin. From the spring of 1940 until early 1941, Heatley was able to improve on the extraction process by fashioning out of glass tubing an automated countercurrent apparatus. Once the filtrate was collected it was acidified and then extracted with ether and then distilled. With the countercurrent apparatus 12 liters of "mold juice" could be processed in an hour; the final product was freeze dried to give a very nice brown powder with undiminished activity. The powder could be diluted a million-fold, and although quite impure it was still able to inhibit bacterial growth.

Edward Abraham, another biochemist who was employed to step up the production of penicillin, used a newly developed technique of alumina column chromatography to remove the chemical impurities. When the filtrate was poured through the column four colored bands were seen; the top brownish-orange band had little penicillin activity, the second pale yellow band was 80% penicillin and free of contaminants that could cause fever, the third orange layer had some or all of the fever causing activity, and the bottom brownish layer was full of impurities and had no penicillin activity. The yellow band was removed, washed in a neutral pH solution, and when diluted exhibited a pale yellow color with a faint smell and a bitter taste. Eventually, this was used to produce crystals for X-ray crystallography. In 1945, Chain and Abraham in collaboration with the X-ray crystallographer Dorothy Hodgkin, also at Oxford University, firmly established the structure of penicillin: a four-membered highly labile beta-lactam ring fused to a thiazolidine ring.

In the meantime, animal tests in mice showed that the penicillin was nontoxic. When a mouse was injected intravenously with the penicillin, it passed through the kidneys into the urine; a drop of urine using Heatley's well diffusion assay showed the penicillin still had antibacterial activity. Chain said, "We knew that we had a substance that was nontoxic and not destroyed in the body and therefore was certain to act against bacteria in vivo." Now, penicillin looked like a drug.

On Saturday May 25, 1940, there was enough penicillin, still less than 1% pure, to determine whether it could protect mice from an otherwise lethal bacterial infection. Eight mice were infected with virulent streptococci and an hour later, four of these were given the crude penicillin. All four untreated mice died within 16 h; all treated mice were alive and well the next day. Florey's remark, "It looks promising" was a typically laconic assessment of one of the most important experiments in medical history. The results of a large series of in vitro experiments with bacteria that cause gonorrhea, boils, diphtheria, anthrax, lobar pneumonia, tetanus, gas gangrene, and meningitis were published in the journal *Lancet* on August 24, 1940, in a two-page article "Penicillin as a Chemotherapeutic Agent."

Florey tried to persuade British drug firms to produce enough penicillin to treat human infections; however, they were already hard pressed with other wartime needs. So he turned his own department into a small factory. Here again Heatley's technical ingenuity came into play. He was responsible for the design of ceramic bed pans in which the *Penicillium* was grown using penicillin girls specially employed for the purpose of preparing medium, adding spores, and decanting media; they were paid 15 shillings a week and after a month their salary was raised 5 shillings to a pound Sterling.

By January 1941, there was enough penicillin for a limited trial on a patient. The first case was a 43-year-old retired policeman, Albert Alexander, who was near death's door from an overwhelming streptococcal and staphylococcal infection that resulted from a scratch while pruning a rose bush. He was given 200 mg of penicillin followed by three doses of 100 mg every 3 h, and made a dramatic improvement over 24 h and after 8 injections his temperature was reduced, pus stopped oozing from the sores, but after 10 days the entire supply of penicillin (~4.4 g or 200,000 units) was exhausted and he relapsed and died. Subsequently, six other similar hopeless cases were treated by intravenous injection of penicillin; in all cases the results were conclusive: penicillin was able to overcome infections that ordinarily would be fatal. The results were published in the August 1941 issue of the *Lancet*.

In the summer of 1941, Heatley and Florey, with the support of the Rockefeller Foundation, traveled to the United States to see if they could interest U.S. pharmaceutical companies in producing penicillin on a large scale. Through the efforts and contacts of John Fulton at Yale University they were put in contact with the Northern Regional Research Laboratory of the Department of Agriculture in Peoria, Illinois. By adding lactose instead of sucrose the mold cultures produced more penicillin and then using corn steep liquor, a product of the corn wet-milling process, the yields were further increased. It was recognized that the Oxford method (using Heatley's ceramic bedpans) of growing the mold on the surface of the nutrient medium was inefficient and growth in a submerged culture would be superior. (In submerged cultures the mold is grown in large tanks and is constantly agitated and aerated.) Florey's isolate of *Penicillium* did not grow well in submerged culture so a more productive strain was sought. It came from a moldy (*Penicillium chrysogenum*) cantaloupe found in a Peoria market. In the United States, the resultant purified penicillin product was named penicillin G.

To promote the development of penicillin, the U.S. government encouraged drug companies to collaborate without fear of potential antitrust violations. By February 1942, Merck and E. R. Squibb & Sons had signed an agreement to share research and production information. They also signed a joint ownership of their inventions to include other firms that made definite contributions and in September Charles Pfizer and Company joined the group. Later, Abbott Laboratories and Winthrop Chemical Company joined them. Eventually there were 21 other pharmaceutical companies. In the first 5 months of 1943, 400 million units of penicillin were produced in the United States—enough to treat about 180 severe cases; in the following

7 months, 20.5 billion units were produced. By the end of 1943, the production of penicillin was the second highest priority at the War Department. Only the development of the atomic bomb was considered more important. In March 1944, Pfizer opened the first commercial large-scale production plant in Brooklyn, New York. By the end of 1945, Pfizer was making more than half of all the penicillin in the world. Between 1943 and 1945, the price per million units of penicillin—enough to treat one average case—dropped from $200 to $6. By 1949, the annual production in the United States was 133,229 billion units and the price dropped to less than 10 cents per unit.

Aftermath

In 1945, the Nobel Prize in Physiology or Medicine was awarded to Alexander Fleming, Ernst Chain, and Howard Florey for "discovery of penicillin and its curative effect in various infectious diseases." The Nobel committee might have added that penicillin was the ideal nontoxic therapeutic in that it interfered with a specific chemical reaction necessary for the synthesis of the bacterial cell wall.

The committee awarding the Nobel Prize did not recognize the contributions made by Norman Heatley to the development of penicillin as a clinical drug. Heatley continued to work at the Dunn School and in the 1950s he worked with Edward Abraham on cephalosporin, but as he confessed years later, no work ever matched the results with penicillin. He retired in 1976, and received no honorary degrees or awards save for a 1978 appointment to the Order of the British Empire. He died in Oxford at age 92.

The relationship of Ernst Chain with Howard Florey continued to deteriorate and in 1947 Chain went to Italy to give several lectures on penicillin and when in 1948 he was invited to direct a research center at the Instituto Superiore di Sanita in Rome he accepted; he did not return to the Dunn School for 30 years. In 1949, he was elected a fellow of the Royal Society and he received numerous honorary degrees and memberships in learned societies. In 1961, he was made Professor of Biochemistry at Imperial College, London; was knighted in 1969; became a consultant to pharmaceutical companies; and was a valuable supporter of the Weizmann Institute in Israel. In 1971, he built a house in County Mayo, Ireland, where he died in 1979 at age 73.

After the Nobel ceremony Florey returned to his laboratory. He was awarded honorary degrees from more than a dozen universities and honorary memberships in learned scientific societies. In 1944, he was made a Knights Bachelor and in 1960 was honored by the Royal Society of Medicine. Florey followed in the footsteps of his mentor Charles Sherrington when he was elected president of the Royal Society for 5 years, and in 1966, he was made provost of Oxford's Queens College. He was made a life peer to become Baron Florey and was appointed a member of the Order of Merit, one of Britain's greatest honors. Even as his health declined, Florey continued to work until the evening of February 21, 1968, when at age 69 he died from a heart attack. In 1982, a commemorative plaque in his honor was put in the floor of the nave in Westminster Abbey near the one for Charles Darwin.

Florey never wrote his memoirs. If he had he might well have claimed that the greatest beneficiaries of penicillin were not the patients who were cured of their bacterial infections but in actual fact Alexander Fleming. Indeed, without Florey's work Fleming would have gone down as a somewhat eccentric microbiologist. After the Nobel ceremony, Fleming's popularity soared and over the next 10 years he averaged more than one tribute a month and was awarded more than a score of honorary degrees. He was made Lord Rector of Edinburgh University; received more than 50 medals, prizes, and decorations; and was an honorary member of more than 90 professional societies. Fleming retired in 1948, was an Emeritus Professor until 1954, and died suddenly from heart attack on March 11, 1955.

Why did Fleming, who was an undistinguished metropolitan laboratory scientist until age 61, and who Ronald Hare called a "third-rate scientist" become such an international celebrity? It was of course because his name became synonymous with penicillin and the antibiotic revolution. The Fleming Myth as it has been called, was promoted by the publicity machine at St. Mary's and the efforts of Lord Beaverbrook, the press baron, who was a generous benefactor and patron of St. Mary's Hospital and considered Fleming a genius; he felt it is his duty to inform the world of Fleming's achievements so Britain could bask in reflected glory. And, Charles Wilson, later Lord Moran, who was president of the Royal College of Physicians as well as Churchill's wartime physician, was a supporter of Fleming's contribution. Edward Mellanby, Secretary of the MRC, and one of the most powerful men in Britain was also a promoter of Fleming. Mellanby, a virulent anti-Semite, considered Chain a money grubbing Jew and unfit for public acclaim. And, because Florey was a gruff, rough colonial he too, in Mellanby's eyes, did not qualify. Fleming on the other hand was the perfect hero, a modest, quiet Scotsman. For his part Fleming did not make unreasonable claims about what he had discovered but neither did he go out of his way to contradict some of the grossly exaggerated claims made by St. Mary's in the popular press.

Florey was a no nonsense scientist who received far less public acclaim than Fleming and had little regard for Fleming or self-promotion. Henry Harris who worked with Florey at Oxford University and succeeded him as Professor of Pathology has said: "Florey was a practical scientist, not a great one. He was not a seer, or a conceptualist, he was no Darwin or Pasteur or Einstein or Ehrlich and he never thought of himself in that league … he had one supreme virtue: he knew what had to be done and he got it done." Where Fleming gave up quickly Florey persisted and came up with solutions. Florey, unlike the loner Fleming, was a team player and a great leader who despite his temperament of sometimes throwing off casual remarks without considering the seriousness, with which others took them, inspired the members of his team to follow his example.

In contrast to drug development today with a cost of hundreds of millions of dollars to bring a drug to market, the Oxford team developed penicillin for a few thousand dollars. The pity of it all is not so much the total cost of antibiotic development, it is the lack of financial incentive on the part of pharmaceutical companies to develop them since antibiotics are usually only good for a few years before the bacteria become resistant.

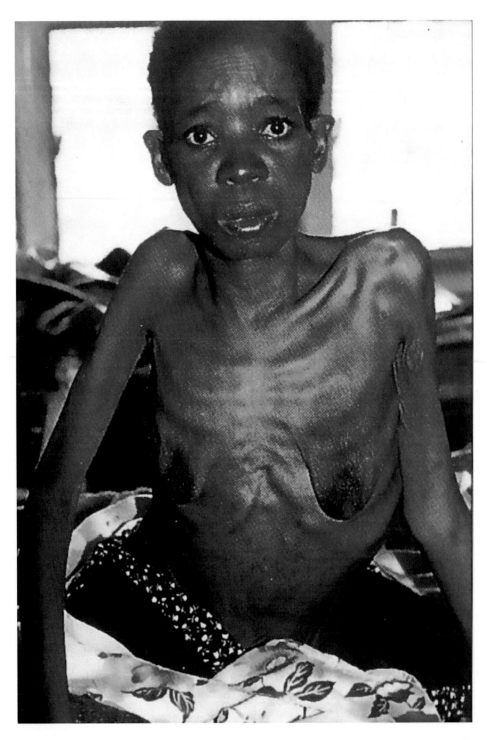

A 25-year-old AIDS patient (also called slim disease in Africa). (Courtesy of Sebastian Lucas.)

Chapter 10

AIDS, HIV, and Antiretrovirals

AIDS (acquired immune deficiency syndrome), called by some the plague of the twenty-first century, is a deadly global disease for which there is no vaccine or cure. The magnitude of the AIDS epidemic can best be appreciated by looking at the increase in the number of individuals infected. In 1981, when the disease first appeared in the United States there were 12 cases, by 1994, the number of cases was 400,000, and in 2015 the CDC estimated that 1.2 million people aged 13 years and older were living with the infection. Worldwide, UNAIDS estimates that in 2014 over 36.9 million people were infected and in that year alone 1.2 million people died of AIDS-related illnesses. It has been claimed that since the epidemic began, 25.3 million people have died from AIDS-related illnesses. With preventive measures and drug treatment, the vast majority of these deaths would not have occurred.

Cause

AIDS is caused by the human immunodeficiency virus (HIV). The discovery of HIV began more than a century ago (1884) with the development of a porcelain filter by Charles Chamberland, who at the time was working in the laboratory of Louis Pasteur. Chamberland's filter had very small pores and it was possible by using this filter, as Chamberland wrote "to have one's own pure spring water at home, simply by passing the water through the filter and removing the microbes." In 1911, a sick chicken was brought to a young physician, Peyton Rous (1879–1970), who was working at the Rockefeller Institute in New York City. The chicken had a large and disgusting tumor in its breast muscle. Rous wondered what could cause such a tumor (called a sarcoma), and so he took the tumor tissue, ground it with sterile sand, suspended it in a salt solution, shook it, centrifuged it to remove the sand and large particles, and filtered it through a Chamberland filter. The sap (or filtrate) was used to inoculate chickens and several developed sarcomas in a few weeks. Rous examined the filtrate and the sarcoma with the light microscope and found neither contained bacteria; he concluded he had discovered an infectious agent capable of causing tumors, that the infective agent was smaller than a bacterium, and that in all probability it was a virus. Rous failed to find similar virus-causing cancers in mice or humans and received no support for his beliefs from other scientists, so the sarcoma story was regarded as a biological curiosity. As a result, Rous turned his attention to other aspects of pathology. Ironically, just 4 years before his death at age 91, Rous's 1911 work was recognized when he received the 1966 Nobel Prize.

In the 1950s, it became possible to grow a variety of cells in laboratory dishes and with these tissue cultures it became possible to study the effect of viruses on isolated living cells rather than in mice or chickens. At this time, it was also found that viruses were of two kinds, those that contain DNA and those that contain RNA. Renatto Delbecco (1914–2012), who began to study DNA viruses in tissue cultures, found that although some viruses caused tumors, on occasion, virus could not be detected in the infected cell because its genetic material (DNA) had become inserted or integrated into the cell's chromosomes. In other words, the virus had been incorporated into the host cell genes and behaved as if it were a part of the cell's genetic apparatus. The virus DNA was now a part of the cell's heredity!

Delbecco worked with DNA viruses, and this made it easier to visualize how the DNA of a virus could be integrated into the DNA of the host cell. However, Rous sarcoma virus (RSV) was an RNA-containing virus, so it was not obvious how the genetic information of RSV could become a part of the tumor cell's heredity. Howard Temin and David Baltimore provided the answer. Temin (1934–1994) attended Swarthmore College majoring in biology and then went on to CalTech (California Institute of Technology) to do graduate work with Renatto Delbecco; his doctoral thesis was on RSV. Temin's experiments, carried out from 1960 to 1964 at the University of Wisconsin, convinced him that when the RSV nucleic acid was incorporated into the genetic material of the host cell, it acted as a "provirus." Under appropriate conditions, the provirus would trigger the cell to become cancerous.

A fellow graduate of Swarthmore College, David Baltimore (1938–), did his doctoral work at MIT and at the Rockefeller Institute where he studied the virus-specific enzymes of RSV. His first independent position was at the Salk Institute in La Jolla, California, with Renatto Delbecco studying RSV in tissue culture. When he returned to MIT as a faculty member he continued to study the RSV enzymes. In 1970, Temin and Baltimore simultaneously showed that a specific enzyme, reverse transcriptase, in RSV was able to make a DNA copy from the virus' RNA. They independently went on to show that the replication of RNA viruses involves the transfer of information from the viral DNA copy, and that the viral DNA was integrated into the chromosomes of the transformed cancer cells. Later, other investigators were able to show that purified DNA from the transformed cancer cell, when introduced into normal cells, caused the production of new RNA tumor viruses. Clearly, RSV was a special kind of cancer-causing virus. For their discoveries Delbecco, Temin, and Baltimore shared the 1975 Nobel Prize for Physiology or Medicine.

This was the setting when AIDS appeared in 1981. Two laboratories—one in France headed by Luc Montagnier and one at the National Institutes of Health (NIH) in the United States headed by Robert Gallo, identified a virus named by Gallo HLTV III (human lymphotropic T-cell virus) or LAV (*lymph*adenopathy *v*irus) by Montagnier in tissues obtained from patients with AIDS. Today both are recognized to be the same virus. The virus was renamed human immunodeficiency virus (HIV), and the disease complex it produced was called acquired immunodeficiency syndrome (AIDS).

Viruses are unable to replicate themselves without taking over the machinery of a living cell, and in this sense, they are the ultimate in parasites. The material containing the viral instructions for replication, that is its genes, may be composed of one of two kinds of nucleic acid, DNA (deoxyribonucleic acid) or RNA (ribonucleic acid), and the viral nucleic acid is packaged within a protein wrapper called the core, this in turn, is encased in an outer virus coat or capsid, and the outermost layer is called the envelope. In the case of HIV, the genetic material is in the form of RNA, not DNA, so in order for this virus to use the machinery of the host cell (which can only copy from DNA), it must subvert the cell's machinery to copying viral RNA into DNA. As a consequence, in HIV the information flows: RNA→DNA→RNA→protein, and because the flow of information appears to be the reverse of what is typically found in cells (DNA→RNA→protein) these viruses (which includes RSV) are called retroviruses.

The Genes of HIV

The genes of HIV code for at least nine proteins. These proteins are divided into three classes: structural, regulatory, and accessory. The major structural proteins are gag (group-specific antigen), pol (polymerase), and env (envelope). The regulatory proteins are Tat and Rev, and the accessory proteins are Vpu, Vpr, Vif, and Nef. The gag gene encodes a long polyprotein that is cleaved by the viral protease to form the capsid. The pol-encoded enzymes are also initially synthesized as part of a large polyprotein precursor; the individual pol-encoded enzymes, protease, reverse transcriptase, and integrase are cleaved from the polyprotein by the viral protease. The envelope glycoproteins are also synthesized as a polyprotein precursor. Unlike the gag and pol precursors, which are cleaved by the viral protease, the envelope precursor, known as gp160, is processed by a host cell protease during virus trafficking to the host cell surface; gp160 processing results in the generation of the surface envelope glycoprotein, gp120, and a transmembrane glycoprotein, gp41.

HIV Targets Immune Cells

In order to understand how HIV is able to cause AIDS it is necessary to digress briefly to discuss the immune system. Blood consists of cells suspended in a yellowish, salt- and protein-containing fluid called plasma (after the blood has clotted the fluid is called serum). There are two kinds of cells in our blood: red blood cells and white blood cells. The red cells contain hemoglobin (responsible for the red color of blood) and are involved in the transport of oxygen to and removal of carbon dioxide from the tissues. The white blood cells have a different function; they are a part of the body's defense system and play a key role in immunity. In our body there are about a trillion white cells and they are of five kinds. There are three kinds of granule-bearing white cells called eosinophils, basophils, and neutrophils, and there are two kinds that lack granules in the cytoplasm, called lymphocytes and monocytes (macrophages). The lymphocytes and macrophages are produced in the bone marrow, but are found in regional centers such as the spleen and lymph nodes as well as in the blood.

Lymphocytes are divided into two different types called T and B. The B-lymphocytes make antibody either on their own or by being activated by a T-helper cell, called T_4. The T_4 lymphocytes are so-named because they have on their surface a receptor molecule, CD4. Macrophages also have CD4 on their surface. T cells, unlike B cells, do not make antibody but they are involved in what is referred to as cell-mediated immunity. Cell-mediated immunity is the immunity that is responsible for transplant rejection, and delayed hypersensitivity reactions such as those that occur after being exposed to a bee sting. The T cells do not make antibody but they can communicate with one another using soluble chemicals called chemokines. Chemokines are soluble chemical messengers that attract or activate other white cells, especially T and B lymphocytes and macrophages. To be activated, these white cells must have on their surface a receptor (analogous to a docking station) for the chemokine. One chemokine receptor called CCR5 is an entry cofactor for the invasion of T cells by HIV and it also activates neutrophils.

Another chemokine receptor called CXCR4 is an entry factor for the invasion of macrophages by HIV and it activates monocytes, lymphocytes, basophils, and eosinophils.

Let's now look more closely at how HIV and immune cells interact with one another. HIV, a roughly spherical particle 1/60th the size of a red blood cell, consists of an outermost layer, the envelope, and within there is an associated matrix enclosing a capsid which itself encloses two copies of the single-stranded RNA genome and several enzymes. The glycoproteins of the HIV envelope resemble lollipops: the "stick" is called gp41 and the "candy ball" is gp120. The gp120 contains the determinants that interact with the CD4 receptor and co-receptor CCR5/CXCR4, while gp41 not only anchors the gp120/gp41 complex in the membrane, it also contains domains that are critical for the membrane fusion reaction between viral and host cell lipid bilayers during virus entry.

The two viral proteins, gp40/gp120, act in concert to anchor the HIV to CD4 on the surface of the T cell. Within an hour of docking, the gp120 changes its shape so that it can bind to the chemokine receptors, CCR5/CXCR4, and this allows the virus to fuse with and to gain entry into the cell. Precisely how virus–host cell membrane fusion occurs is not well understood but once HIV enters the cell the capsid breaks down and the viral RNA and reverse transcriptase are released. The latter begins to synthesize viral DNA from the viral RNA; the pro-viral DNA moves to the nucleus of the cell where another viral enzyme, called integrase, inserts the pro-viral DNA into the DNA of the cell. When this infected cell begins to make proteins it synthesizes (via viral mRNA) a long string of polyprotein. After this, a viral enzyme, protease, cuts the polyprotein into smaller proteins that serve a variety of viral functions: some become structural elements (envelope, capsid) while others become enzymes such as reverse transcriptase, integrase, and protease. Once this process (which takes about 15 h) has been completed, the newly assembled virus particles move to the inner surface of the membrane of the host cell where they mature. Following maturation, the infectious viral particles are released into the bloodstream by budding from the surface of the HIV-infected cell. The entire process from virus binding to virus release takes ~48 h and during that time a single virus particle can produce several hundred thousand new infectious particles. And, in an infected individual every day 10.3 billion virus particles can be produced. At present, there are approximately 30 approved therapeutic drugs targeting the various steps in the replication cycle: (1) entry, (2) reverse transcription, (3) integration, (4) maturation, and (5) protein cutting.

AZT: The First Antiretroviral Drug

When Jerome Horwitz (1919–2012), an organic chemist at the Michigan Cancer Center, Wayne State University, Detroit, Michigan, opened the March 27, 1987 issue of the *New York Times* and read the headline "U.S. Approves Drug to Prolong Lives of AIDS Patients," he was stunned. The reason for Horwitz's astonishment was that the "approved drug"—AZT or azidothymidine—was a compound he had spent a decade developing as a cure for leukemia, however, in the 1987 article, AZT was named as an antiviral drug made by Burroughs Wellcome (now GlaxoSmithKline, GSK) under the brand name Retrovir.

According to the article, AZT had been tested on AIDS-infected individuals in the United States as early as July 1985, with a pivotal clinical trial beginning in February 1986. During the clinical trial 19 deaths occurred among 137 patients taking the placebo compared to a single death among the 144 AIDS patients taking AZT. By the time of its approval by the FDA, AZT had been made available to more than 5000 patients with AIDS. AZT reached the market with remarkable speed proceeding from laboratory to clinical trails and FDA approval in less than two-and-a-half years. Horwitz's contribution to the development of AZT was never mentioned.

In the early 1960s with the war on cancer in full swing (and before there was any thought of an AIDS epidemic), the approach by many scientists was to randomly pull drugs off the shelf to test whether they had any potential as anticancer agents, but Horwitz found this approach intellectually unsatisfying. Instead, he decided on a more rational approach. Knowing that dividing cells, especially cancer cells, require nucleosides (for the synthesis of DNA) he created what he called "fraudulent nucleosides" that were similar to the real thing. These "fakes" he theorized would gum-up the cancer cell's replication machinery to halt its rapid division. The theory was fine, however, when AZT (3'-azido-2', 3' dideoxythymidine) was synthesized in 1964 and it failed to help mice with leukemia he "dumped it on the junkpile," wrote up his failure and moved on. He was so disappointed in AZT that he did not even apply for a patent.

AZT collected dust on the shelf until the mid-1980s when there was a public awareness of the growing number of deaths from AIDS. This prompted a search for a treatment. At Burroughs Wellcome, as in many pharmaceutical company laboratories, the research goal was to find inhibitors for particular biological targets. Researchers at Wellcome had experience with drugs against herpes viruses and other viruses and they began testing compounds that might be effective against retroviruses. AZT showed promise in laboratory animals and in collaboration with the National Cancer Institute (NCI) and in discussion with the FDA Wellcome sought permission to begin the testing of AZT in humans. On July 3, 1985, the first AIDS patient received AZT at the NCI and the result was encouraging. Further studies in 19 patients with AIDS showed that the drug harmed the bone marrow at high doses but it was not seriously toxic; more importantly many of the AZT treated individuals improved. Wellcome applied for and received a patent for AZT and on March 20, 1987, the FDA approved AZT for the treatment of HIV infections. Marketed as Retrovir by Wellcome it was priced at $8000 a year. By 1992, AZT had sales of $400 million. Without a patent for AZT, Horwitz did not receive a penny, although Wellcome did donate $100,000 to the Michigan Cancer Center in Horwitz's name but it was not enough to cover the costs of an endowed professorship. Horwitz remained at Wayne State, developed two anticancer drugs (2', 3' dideoxycytidine and 2', 3'dideoxy adenosine), patented them, and in 2003, Wayne State University licensed them to a pharmaceutical company. The company paid a hefty licensing fee, and at age 86, Horwitz received the first royalty check of his career. Horwitz died in 2012 at age 93.

In 1942, George Hitchings at the Burroughs Wellcome Company laboratories in Tuckahoe, New York, began to study analogs to inhibit the growth of malignant cells and malaria (see p. 133). The underlying principle was that even if there was not an absolute difference in the metabolism of the malignant cell, there were probably enough differences to allow for differential inhibition. In 1944, Gertrude Elion (1918–1999) joined Hitchings at Wellcome. Although at the time, none of the enzymes or the steps in the formation of nucleic acids were known and the deciphering of the double helix structure of DNA by Watson and Crick was a decade away, Hitchings and Elion began studying the growth inhibitory properties of analogs of pyrimidines and purines. And 44 years later, Hitchings and Elion shared the Nobel Prize for Physiology or Medicine "for ... discoveries of important principles for drug development."

Gertrude Elion was born in New York City to parents who immigrated to the United States from Russia and Lithuania. At age 15, she was motivated to embark on a research career when her beloved grandfather died from stomach cancer. Elion's parents were poor so she attended the free college, Hunter College, with a major in chemistry. Graduating in 1937, and with the world-wide economic depression, she was unable to go directly to graduate school so she worked as a teacher, and a laboratory technician. Two years later, with some money saved and with the help of her parents, she enrolled for graduate study at NYU and was the only female in the chemistry classes. During the time she was enrolled for an MS degree, she taught chemistry in New York secondary schools by day and worked on research at night graduating in 1941. With the ongoing World War II, jobs began to open up for women in chemistry. Her first chemistry job however was far from glamorous: testing the acidity of pickles and the color of mayonnaise for the A & P grocery chain; later she was hired by Johnson & Johnson to synthesize sulfa drugs, but the lab closed after 6 months. In 1944, she was hired by Hitchings and began to study for the PhD at the Brooklyn Polytechnic Institute taking evening courses; however, when she was told that she would have to become a full-time student to complete the PhD she abandoned that goal. She remained at Wellcome for the rest of her career and by the time she retired, 39 years later, she was the holder of 45 patents, had 23 honorary degrees, a long list of honors, and was the head of the Department of Experimental Therapeutics. Despite receiving the Nobel Prize she never completed the requirements for a PhD!

Elion's initial job was to make purine compounds that would antagonize the growth of the bacterium *Lactobacillus*. To do this she went to the library, looked up the methods for synthesis found in the old German literature, and made the compounds. In 1948, she had in hand a 2,6 diamino purine that inhibited the growth of *Lactobacillus*, and it showed in vitro activity against the DNA vaccinia virus, however, it was too toxic for animals and so was abandoned. By 1951 Elion had made and tested over 100 purine analogs in the *Lactobacillus* screen and discovered that two—6-mercaptopurine (6-MP) and 6-thioguanosine (6-TG) were also particularly effective against rodent tumors and mouse leukemia. The finding that 6-MP could produce complete remission of acute leukemia in children (although most relapsed later) led to its approval by the FDA in 1953. With combination therapy, using 3 or 4 drugs to produce and consolidate remission, plus several years of maintenance therapy with 6-MP and methotrexate 80% of children with acute leukemia were cured. In 1959, the compound azathioprine (marketed as Imuran) was found to blunt transplant rejection especially in kidney transplantation. And in 1968 Elion turned her attention to antivirals. Diamino arabinoside was found to be active against herpes simplex and vaccinia virus, and in 1970, the guanine analog, acycloguanosine or acyclovir, was found to be effective against herpes, varicella, and Epstein–Barr virus but not cytomegalovirus. Acyclovir, marketed as Zovirax, is not toxic to mammalian cells; it interferes with the replication of herpes virus and only herpes virus proving that these drugs can be selective. In time the discoveries of acyclovir, 6-MP and 6-TG would lead to the development of AZT by Elion's colleagues at Wellcome.

Antiretroviral Drugs

Before the mid-1990s, there were few antiretroviral treatments for HIV and the clinical management of those with AIDS consisted largely of prophylaxis for common opportunistic infections. In 1987, AZT was the first antiretroviral used in the treatment of AIDS. AZT acts by targeting the HIV reverse transcriptase and belongs in the category of nucleoside reverse transcriptase inhibitors (NRTI). Although AZT blocks the activity of the HIV reverse transcriptase, and the drug has a 100 to 1 affinity for HIV over human cells, it does limit the polymerase needed by healthy cells for cell division when used at high doses. And at higher concentrations it depresses the production of red and white blood cells in the bone marrow.

Following AZT (zidovudine) other NRTI were developed: abacavir, tanofovir (an analog of AMP), lamivudine, diadanosine (ddI), zalcitabine, stavudine, and emtricitabine. Treatment of HIV was further revolutionized in the 1990s with the introduction of these NRTI, as well as the nonnucleoside reverse transcriptase inhibitors (NNRT) efavirenz, nevirapine, rilpivirine and the protease inhibitors: lopinavir, idinavir, nelfinavir, tripranavir, saquinavir, darunavir, amprenavir, fosamprenavir, atazanavir, and ritonavir. This allowed for FDA approval of regimens consisting of combinations of antiretroviral drugs, called highly active antiretroviral therapy or HAART. (HAART is initiated when the T lymphocyte count declines to 350/mm^3 and when there are between 10,000 and 100,000 HIV particles per ml of plasma.) In addition, the HIV integrase was recognized as a potential antiretroviral target as early as 1995 when susceptibility was found by oligonucleotides, synthetic peptides and polyphenols, but the milestone was the publication in 2000 of a diketoacid inhibitor that led to FDA approval of the first integrase inhibitor: raltegravir (approved in 2007); following this, was the clinical development of elvitegravir (approved in 2012) and dolutegravir (approved in 2013). There is also a CCR5 antagonist maraviroc available for HIV therapy.

Currently, there are 15 million people taking antiretroviral drugs and it is estimated that 30 million infections have been averted between 2000 and 2014 by HAART. HAART can suppress viral replication for decades; however, it alone cannot eliminate all HIV particles. Nevertheless, HAART has changed HIV from an inexorably fatal condition into a chronic disease with a longer life expectancy. Indeed, in 2001 the life expectancy of an individual with HIV was 36 years, in 2014 thanks to HAART it was 55 years!

In 2001, HIV treatment required 8 or more antiretroviral pills a day at a cost of $10,000 per year, but by 2014 a variety of single daily tablet regimens (tenofovir/emtricitabine/efavirenz or tenofovir/emtricitabine/rilpivirine or tenofovir/emcitricitabine/elvitegravir or abacavir/lamivudine/dolutegravir) were introduced at an annual cost of $100.

As with most viruses, the AIDS-causing virus is hard to kill without doing any damage to the host. And even with combination drug therapy resistant viruses may emerge. Treatment by an antiretroviral drug can be a two-edged sword: it may benefit the individual, but if it does not reduce infectiousness it might not significantly benefit the community. It has been calculated that the number of virus particles must be reduced to less than 50 per μL of serum for the individual to lose the capacity to be infectious. (However, it has been shown that even with successful reduction a viral reservoir may remain in the body and if antiretroviral therapy is discontinued or if there is the emergence of resistance the virus may rebound.)

The Nature of Drug Resistance

Perhaps the earliest notion that tolerance to a drug could result from prolonged use was that of the Turkish King Mithiridates (119–63 BC), who "in order to protect himself from the effects of poison had the habit of taking such substances prophylactically with the result that, when ultimately defeated in battle by Pompey and in danger of capture by Rome, was unable to take his own life by poison and was obliged to perish by the sword." Although known since ancient times that the effectiveness of narcotic drugs such as morphine gradually lost their effects by repeated usage, the recognition of the problem of drug resistance by microbes and insects first began with Paul Ehrlich's use of "magic bullets." Working with mice infected with trypanosomes (causing the cattle disease nagana) he gave the mice a red aniline dye (trypan red) that was curative, that is, the parasites disappeared from the blood, however, shortly thereafter they reappeared. Further treatment of infected mice showed the dye to be ineffective with mice dying rapidly. The dye-treated strain of trypanosomes when inoculated into a batch of normal mice produced infections that even in the absence of drug were tolerant to the killing power of the dye suggesting that a genetic change in the parasite resulted in their renewed strength.

What is the basis of resistance? Just as there is resiliency in our species to adapt to new environmental challenges there is also a genetic resiliency in other species that enable them to survive the onslaught of a drug. Oftentimes, resiliency lies in favorable mutations that permit an organism to survive an environmental threat. This capacity is then passed on to its offspring, that is, survival of the fittest. The presence of the drug (or insecticide) acts as a selective agent— a sieve if you will—that culls out those that are susceptible and allows only the resistant ones to pass through to the next generation.

Resistance to a drug is the result of natural selection that is, those with a particular genetic makeup (genotype) that are the most able to survive and reproduce their kind pass on their genes to future generations and increase in frequency over time. Drug resistance could develop in this way: let's assume that an average HIV patient has 10,000–100,000 virus particles per ml of plasma. Owing to the error-prone reverse transcription process (which takes about 10 h) it is estimated there is one mutation for every 1,000–10,000 nucleotides synthesized. Since HIV has ~10,000 nucleotides with each replication cycle 1 to 10 mutations may be generated. Thus, there is an enormous potential for the appearance of genetic variants with reduced susceptibility to one or two drugs. Let's assume that an HIV infected individual has a billion virus particles and 1 in 10,000 of these carries a mutation that allows that virus to evade the lethal effects of a retroviral drug. Once the patient is treated with a single drug only the mutant survives and these are drug resistant. The result is that there are now ~10,000 virus particles and these can reproduce in the presence of the drug, increase their numbers, and now almost the entire population is resistant. But let's assume that a second, and equally effective, antiretroviral drug is added along with the first drug to the billion viruses. Again, only 1 in 10,000 is resistant to this drug. When the drugs are added together there is less than 1 resistant virus. By adding a third drug the possibility for a surviving virus is reduced even further. Thus, HAART decreases the probability of selecting viruses that have multiple mutations conferring resistance to a regimen involving three antiretroviral drugs.

Failure to Control by Antibody or Vaccine

Soon after the identification of HIV, a specific and sensitive test for antibodies to HIV was developed. Using such a diagnostic test it has been possible to screen the serum of large numbers of individuals and to determine those who are infected. (Such individuals are called serum positive or seropositive.) This simple test also had immediate and profound effects for public health since now blood supplies in the United States and other countries could be screened for HIV. By 1985, through the use of screening, it was possible to ensure that the blood supply was HIV-free, thereby preventing millions of potential transfusion-related infections. In addition, the antibody test could be used in epidemiologic studies to determine the global scope and evolution of the disease. The availability of the antibody test has allowed for the identification of individuals before they show clinical signs of the disease and permits a more accurate description of the true clinical course of HIV infections. The FDA approved a rapid HIV antibody test that can be performed outside the laboratory and provide results in about 20 minutes. Those with positive results should have a confirmatory test, however, those that are seronegative should recognize that although they are presently without evidence of infection in time they could become positive and therefore a repeat of the test would be necessary to ensure the individual is HIV-free.

Although antibodies to HIV can be found in the plasma of millions of people worldwide these individuals continue to be infected and may transmit the virus to others. Clearly, although HIV elicits a strong antibody response these antibodies neither neutralize the virus nor do they clear it from the body, and there is no protection against reinfection. Why? The antibodies to HIV are unable to reach the critical receptor sites and do not block the conformational changes in gp120 that are necessary for HIV attachment and entry. Further, HIV is able to persist for long periods of time in nondividing T cells so that even if an effective antibody were produced its large size would not allow it to enter the infected cell and kill the virus. In addition to its ability to hide within resting T cells, HIV is able to cripple the immune system by depleting the helper T cells not only by outright destruction (through viral release) but also by decreasing the rate of production of T helper cells. The result can be the appearance of deadly opportunistic infections that may include candidiasis of bronchi; invasive cervical cancer; coccidiomycosis; cryptococcosis; cryptosporidiosis; cytomegalovirus disease (particularly CMV retinitis);

encephalopathy, Herpes simplex; histoplasmosis; isosporiasis, Kaposi's sarcoma; lymphoma; *Mycobacterium avium* complex; tuberculosis; *Pneumocystis carinii* pneumonia; pneumonia, recurrent; progressive multifocal leukoencephalopathy; *Salmonella* septicemia; and toxoplasmosis of brain.

Despite decades of attempts to produce a vaccine for HIV there is none. The reason there's no vaccine for HIV/AIDS, the National Institutes of Health explains, is because HIV has "unique ways of evading the immune system, and the human body seems incapable of mounting an effective immune response against it." A recently published study (*Proceedings of the National Academy of Sciences*, US 112 [3]: 518–523, 2015) using monkeys further clarifies the problem. Five different strategies for immunizing 36 rhesus monkeys against the simian immunodeficiency virus, SIV, were evaluated. After being given an initial shot of one of the five different vaccines, each of the monkeys received booster shots at 16 weeks and then again at 32 weeks. Next, the monkeys were exposed to a low dose of SIV. In general, the researchers found none of the vaccines prevented an SIV infection. Oddly, all the immunized monkeys had detectable levels of circulating "killer" T cells but these cells did not prevent infection. "The possibility that certain immunization regimens designed to protect against HIV infection and AIDS result in increased risk of virus transmission is not just a theoretical concern, because three recent large-scale clinical trials … have shown a trend toward higher infection rates in vaccinated individuals than in placebo recipients," noted the authors.

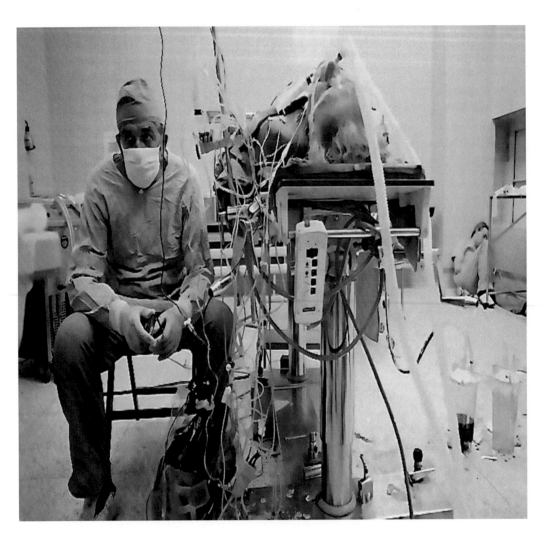

A 1987 photo of Dr. Zbigniew Religa after a 23-hour heart transplant operation. The exhausted surgeon is keeping watch on the vital signs of the patient, and in the lower right corner you can see one of his colleagues who helped has fallen asleep. (Courtesy of National Geographic.)

Chapter 11

Organ Transplantation and Cyclosporine

"I am a new Frankenstein," Louis Washkansky told the nurse at his bedside. The macabre humor was appropriate: Washkansky, a sturdy 53-year-old grocer from Cape Town, South Africa, had just received the heart of a 25-year-old girl who had died in a traffic accident. Washkansky had literally been brought back to life by a remarkable surgical technique performed by Dr. Christiaan N. Bernard (1922–2001) on December 3, 1967. However, the success was short-lived because the immunosuppressive drugs used to keep Washkansky's body from rejecting the new organ had so weakened him that he died from pneumonia 18 days after the operation.

Transplant surgery did not exist before the mid-1960s and, its possibility was derided by many medical scientists simply because the immunosuppression protocols needed for preventing the patient's immune system from turning against the new organ were so poor. Indeed, at the time, the large doses of immunosuppressive drugs or whole body x-irradiation to destroy the immune system that was necessary to prevent organ rejection left the patient vulnerable to deadly infections.

In 1959, Robert Schwartz and William Dameshek discovered that it was possible to suppress the formation of antibodies when rabbits were given foreign antigens (bovine serum albumin) if the rabbits had been given 6-mercaptopurine (6-MP). Hitchings and Elion developed 6-MP as an anticancer drug because it affects all dividing cells, but it also kills the cells that exert an integral function in rejection so when Barnard used 6-MP for immunosuppression in the heart transplant surgery, the large doses he had given did not allow for lengthy patient survival. Although Barnard's work inspired surgeons around the world to try their skills at performing similar heart transplants, within 2 years more than 60 teams had replaced ailing hearts in some 150 patients, of which 80% of the transplant recipients died in less than a year. Surgeons grew discouraged and by 1970 the number of heart transplants had plunged to 18. Clearly, better means for blunting graft rejection were needed.

In 1950, Roy (now Sir Roy) Calne as a medical student was given personal responsibility for a patient who was dying of kidney failure. The senior consultant told him the patient would be dead in 2 weeks, so all Calne had to do was to make the patient as comfortable as possible. When Calne asked the consultant whether the patient could receive a kidney graft he was told, "It can't be done." Calne was perplexed because only three plumbing junctions—an artery, a vein, and the ureter—were required to "hook up" a kidney, and there were surgical techniques available to accomplish the task. At the time, Calne, ignorant of the phenomenon of graft rejection, had no idea why the consultant had made so dire a prediction. Calne went on to complete his medical training (1952) and then served in the Royal Army Medical Corps from 1953 to 1955. Later, he became a lecturer in anatomy at Oxford University. In 1959, at Oxford after hearing a lecture by Peter Medawar on the immunologic basis of graft rejection, Calne returned to the subject of kidney transplantation. Working at the Royal College of Surgeons in London, he found that total body x-irradiation failed to prolong the survival of a kidney transplant. When he learned of the work of Schwartz and Dameshek, he contacted Gertrude Elion at Burroughs Wellcome for compounds that might be superior (and as effective and less toxic to the bone marrow) than 6-MP. Elion suggested he try azathioprine (marketed by Wellcome as Imuran), a drug that had been synthesized in 1957 to produce 6-MP in a metabolically active but masked form. In 1960, Calne received a Harkness Fellowship to study at the Harvard Medical School. While in Boston, Calne carried out many kidney transplant experiments in dogs at the Peter Bent Brigham Hospital in Boston using azathiopurine. When Lollipop, a dog that survived for 6 months, tail-waggingly bowed to applause in a crowded auditorium, Calne was convinced of the effectiveness of the immunosuppressive protocol using azathiopurine and prednisone. On returning to the United Kingdom in 1961, he continued this work and on April 5, 1962, he was able to successfully transplant the kidney for the first time to a genetically unrelated human recipient. For many years thereafter, this kind of dual therapy with azathiopurine and glucocorticoids was the standard antirejection regimen for organ transplantation.

Cyclosporine

In the 1950s, following hard on the discoveries of penicillin and streptomycin, the Swiss pharmaceutical company Sandoz (now Novartis, Basel, Switzerland) established a program to find new antibiotic drugs from fungi. Sandoz employees on business trips and on vacation were encouraged to take with them plastic bags for collecting soil samples to serve as a source of fungi. At Sandoz, these soil samples would be cataloged and the contained fungi screened for antibiotic activity. In March 1970, the Microbiology Department at Sandoz screened two soil samples, one from Wisconsin and the other from Hardanger Vida in Norway that had been collected by employees. The Wisconsin sample contained *Cylindrocarpon lucidum* and the Norway sample *Tolypocladium inflatum*. From these fungi, Z. L. Kis, a microbiologist, isolated neutral, lipophilic metabolites that were novel cyclic peptides. Only the *Tolypocladium*, however, could be grown in submerged cultures to obtain metabolites; the *Tolypocladium* metabolites were not released into the culture medium but required extraction from the filamentous hyphae. After extraction, Kis found two metabolites each containing 11 amino acids, 10 of which were known and with one unique beta-hydroxy, singly unsaturated amino acid, 4(R)-4-(E) 2-butenyl-4, *N*-dimethyl-L-threonine. One of these metabolites had a very narrow range of activity so it was not pursued further by the Microbiology Department as an antibiotic. The other metabolite, preparation 24–556, was sent to the Pharmacology Group where a wide variety of pharmacologic tests could be carried out.

In 1966, within the Pharmacology Group, headed by Dr. Hartmann F. Stahelin, an immunology laboratory was established under the direction of Sandor Lazary. The aim of this laboratory was to search for immunosuppressive agents without toxicity. For this, Lazary and Stahelin developed a mouse test in which immunosuppression (by measuring inhibition of hemagglutination) and cytotoxicity (using P-815 mastocytoma cells) could be assayed in the same animal after administration of a test compound. In the late 1960s, using this assay the Pharmacology Group had discovered ovalicin, a fungal metabolite, which had antitumor activity, depressed the immune response, and inhibited the swelling of the joints in arthritic rats. In 1969, studies in humans were carried out. Ovalicin inhibited antibody formation, however, because it produced a pronounced decrease in blood platelets in several human volunteers further trials were abandoned.

In May 1970, Jean Borel joined the Pharmacology Group and took over the well-equipped immunology laboratory from Lazary who was leaving the company for an academic position. Borel was born in Antwerp, Belgium, in 1933 and 4 years later moved with his parents to Switzerland. He intended to become an artist and with the help of an excellent art teacher and partly on his own acquired a solid background in fine arts. After successfully passing the final gymnasium exams, it was the custom to undertake a cultural trip. In autumn 1953, at the age of 20, Borel traveled to Florence, Italy, where he spent a dozen unforgettable days. However, returning home he had to face a tough reality. His parents insisted he engage in a sensible career. Grudgingly he enrolled in the Swiss Federal Institute of Technology in Zurich where he chose to study agriculture (and in the process completely gave up drawing and painting for the next 30 years). In 1964, at the same institute he received a PhD in immunogenetics. Following his dissertation, Borel spent a year in Madison, Wisconsin, in the Department of Genetics and then a year at the Texas Agricultural and Mechanical College in the Poultry Science Department. He then took up a post back at his alma mater in Zurich studying the immunogenetics of chickens. Subsequently, he switched his scientific interest to human immunogenetics at the Swiss Red Cross in Berne and in 1965 Borel shifted his interest to immunology and inflammation in the Department of Medicine, Swiss Research Institute in Davos Platz where he remained until 1970; in that year, on the advice of a friend, he left academia and went to work for Sandoz Ltd in Basel. Later, Borel became director of the immunology and microbiology departments at Sandoz, was Professor of Immunopharmacology at the University of Bern, and in 1983 became vice-president of the Pharma division of Novartis.

In December 1971, the fungal metabolite preparation 24–556 was found to have immunosuppressive activity in the hemagglutinin assay originally developed by Stahelin and Lazary and modified by Borel. The assay involved immunizing mice with sheep red blood cells, preparing a solution of the test compound, injecting the mice intraperitoneally with the compound on days 0, 1, 2, and 3 and on day 7 bleeding the mice, obtaining the serum from the mouse blood, and testing the serum for its ability to clump sheep red cells (= hemagglutination). The initial test was done in Stahelin's laboratory and the titration was done by Borel. In the initial test of 24–556 the titer of agglutinating antibodies was reduced by a factor of 1024 in comparison to the controls; the preparation did not inhibit the P-815 mastocytoma cells in vitro and did not prolong the survival of mice with leukemia. When Borel repeated the test, the results showed only a fourfold reduction of hemagglutinin titer. Had the disappointing results of the second test been found initially, it is likely further testing would not have occurred. (It has been suggested that in the first test the 24–556 preparation had been solubilized in DMSO and Tween 80, whereas in the second trial the preparation was suspended without organic solvents and therefore was poorly absorbed.) Despite this, Borel and his group persisted and they also detected suppression of antibody when 24–556 in suspension was given intraperitoneally or orally at high doses. Borel's group also found a beneficial effect on graft versus host disease in mice and rats and found the preparation specifically inhibited lymphocytes from the spleen. In 1973, H. U. Gubler in the Pharmacology Group at Sandoz found that a purified preparation of 24–556 (numbered as 27–400) reduced joint swelling in arthritic rats. In 1975, toxicology studies by the Toxicology Group showed no adverse effects in dogs. By 1976, the chemical structure was determined. Later in 1982, R. Wegener was able to produce the molecule synthetically. The first human trials with 27–400 carried out in 1976 were failures presumably because the compound was given as a powder in capsules and was not absorbed. In early 1977, three different preparations of 27–400 were tried in human volunteers (Stahelin, Borel, and von Graffenreid). Borel took the highly hydrophobic compound, mixed in ethanol, water, and Tween 80. It was a distasteful concoction that Borel said made him tipsy, but 2 h later the best hemagglutinin inhibiting activity was found in his serum.

The first report outside Sandoz on the biological activity of 24–556, and now called cyclosporine (because it was a cyclic peptide found in fungal spores), was given by Borel at a meeting of the British Society for Immunology, and a more detailed description of the biological effects was published in *Agents and Actions* by Borel, Feuer, Gubler, and Stahelin. It was in that publication that selectivity for T lymphocytes (not B lymphocytes) was first suggested. Borel's presentation at the meeting excited clinicians in Great Britain, particularly Calne's group in Cambridge. They requested and successfully tested cyclosporine in transplantation of hearts, kidneys, and bone marrow in rats, dogs, and pigs. In 1978, Calne first used cyclosporine with patients with kidney and bone marrow transplants using parenteral and oral preparations. Further testing in a large number of laboratories and clinics confirmed these early results and Sandoz introduced cyclosporine as a drug (Sandimmune) for organ transplantation. The success of cyclosporine suffered a setback in 1979 when studies showed it to be toxic to the kidney and to cause lymphomas. These side effects proved to be the result of administering very high doses of the drug since at the time it was the practice to administer as much drug as the body could handle, short of a toxic level. Further research showed cyclosporine levels could be reduced— just enough to prevent rejection of the transplanted organ—and this eliminated lymphomas and reduced kidney toxicity. Later transplant research by Thomas Starzl in Colorado indicated that cyclosporine worked most effectively for transplantation when administered in conjunction with steroids. In March 1980, a major advance came when 12 of Starzl's liver transplant patients began taking cyclosporine in combination with prednisone. Of these 12 patients, 11 lived for 1 year or longer. Cyclosporine now became the drug of choice and in November 1983, the U.S. Food and Drug Administration (FDA) approved cyclosporine for transplant rejection.

Who Did What?

Jean Borel's name has come to be associated with the discovery of cyclosporine. In 1983, in an article in *Science* magazine, Borel was named as "the hero of the cyclosporine story who discovered the immunosuppresive effects … and doggedly insisted that it be developed." For his "discovery of cyclosporine" in 1986, Borel received the prestigious Gairdner Foundation International Award, in 1987 the Paul Ehrlich and Ludwig Darmstadter Prize, and in 1991 he received an honorary doctorate from the University of Basel. But, was Borel actually the discoverer of cyclosporine or did he, as Isaac Newton once wrote to his rival Robert Hooke, see further by standing on the shoulder of giants?

In 1973, Borel claimed that Sandoz was reevaluating research goals and anticipating that a very large sum of money, on the order of $250 million, would be needed to pursue cyclosporine for approval by the FDA, yet at the time the potential market for immunosuppressives was small, and there was the possibility that cyclosporine might fail as did ovalicin, so the company considered the possibility of abandoning the drug altogether. Borel has said it was he who championed cyclosporine and proposed that it be used in a Sandoz-approved area of research—inflammation—and that was the reason it was considered for further development. Indeed, earlier tests in rats with experimental allergic encephalitis and arthritis and treated with cyclosporine improved. Moreover unlike other drugs, ulcers were not induced. Consequently, cyclosporine was initially promoted for further development for rheumatoid arthritis. According to Hartmann Stahelin, head of the Pharmacology Group at Sandoz from 1955 to 1979, this claim by Borel may be an exaggeration.

The contention made by Borel that "they (my bosses) ordered me to pour cyclosporine down the drain ... (and) several times I was forbidden to work with it. So I worked in secret" is also disputed by Stahelin, and there are no Sandoz documents to substantiate his claim. Borel wrote, "It so happened that it was I who discovered, in January 1972, the marked immunosuppressive effect of the metabolite 24–556." Again, Stahelin disputes this and notes that only the last step of the crucial experiment was carried out in Borel's laboratory by his technician who recognized the strong immunosuppressive effect and called it "interesting." Borel quotes Pasteur's aphorism "chance favors the prepared mind" suggesting that he had the mind prepared to realize the importance of the formulation for proper absorption, but Stahelin, clearly irked by the plaudits showered on Borel, states that he alerted Borel that the way he was using the drug (as a powder) it would be poorly absorbed and suggested Borel dissolve it in Tween 80, alcohol, and DMSO.

Stahelin should not have been astonished that Borel's name has been put on the same level as Alexander Fleming and the discovery of penicillin because it is clear that the credit accrued to Borel (as well as Fleming) was unbalanced and distorted and, in many cases, Borel's (and Fleming's) role was overemphasized.

So, why has Borel's (and not Stahelin's) name been associated with the "discovery" of cyclosporine? Because Borel established and maintained important contacts with outside investigators who at an early stage realized the importance of the potentially unique contribution of cyclosporine to the development of transplantation medicine; also Borel published and lectured extensively and maintained contact with those scientists outside Sandoz, especially in the later stages of cyclosporine development in which Stahelin was not involved. In 2001, it was concluded by a reviewing body that both Stahelin and Borel played instrumental roles in the discovery of the biological effects of cyclosporine and Stahelin's contribution was equally relevant as that of Borel. However, it should be recognized that in the preclinical development of cyclosporine the discovery was actually made by a multidisciplinary team of scientists. It is because the critical immunological data were provided by Borel, and he played a leading role in stimulating research on cyclosporine by outside investigators that he alone has most often been singled out for recognition.

Organ Transplantation

A sea change in organ transplantation occurred with the arrival of the drug 6-mercaptopurine (6-MP) and its derivative azathioprine. These drugs were tested first in a rabbit skin graft model and subsequently by Roy Y. Calne (b. 1930) in the canine kidney transplant model. Although approximately 95% of the mongrel canine kidney recipients treated with 6-MP or azathioprine died in less than 100 days from rejection or infection, occasional examples were recorded of long-term or seemingly permanent graft acceptance. The number of these animals was discouragingly small, but it was an accomplishment never remotely approached using total body irradiation, with or without adjunct bone marrow injection.

Although Thomas Starzl's (b. 1926) primary interest has been liver transplantation, he has also written 148 articles concerned primarily or exclusively with kidney transplantation. Starzl systematically evaluated azathioprine with the simpler canine kidney model rather than with liver transplantation. The yield of 100-day canine kidney transplant survivors treated with azathioprine was small. However, two crucial findings were clinically relevant. First, kidney rejection developing in the dog under azathioprine invariably could be reversed by the addition of large doses of prednisone. Second, the mean survival of the kidney in dog recipients was doubled when the animals were treated with the drug before as well as after the operation. Clinical trials of kidney and liver transplantation followed, in that order. Daily doses of azathioprine were given for 1 to 2 weeks before as well as after transplantation, with the addition of prednisone only to treat rejection. In 1963, Starzl used 6-MP and azathioprine in the first clinical trials in humans of kidney transplantation. Although the follow-up of the first human renal transplantations was only 6 months, in 2002, nine of these patients still had their original kidney. In the United States, between 1988 and 2015, there were 380,000 kidney transplants and the 3-year graft survival rate after transplantation varied between 83% and 94%. About 98% of people who receive a living-donor kidney transplant live for at least 1 year after their transplant surgery. About 90% live for at least 5 years. About 94% of adults who receive a deceased-donor kidney transplant live for at least 3 years after their transplant surgery, and about 82% live for at least 5 years.

Armed with the early kidney experience, the first attempt at liver transplantation was made by Starzl on March 1, 1963. The patient bled to death during the operation. The next two recipients, both adults, died 22 and 7.5 days after their transplantations on May 5 and June 3, 1963. During the last half of 1963, four more failed attempts to replace the human liver were made: two in Denver, and one each in Boston and Paris. Clinical activity then ceased for three-and-a-half years. When the liver program reopened in July 1967, multiple examples of prolonged human liver recipient survival were produced, using triple drug immunosuppression: azathioprine, prednisone, and antilymphocyte globulin.

The opening of Roy Calne's clinical program in Cambridge, England, in February 1968, reinforced the liver transplant beachhead. Despite the successes, the widespread use of the liver and other extrarenal organs, and even of cadaveric kidneys, was precluded for another decade by the high patient mortality. By the end of the twentieth century, however, transplantation of the liver had become an integral part of sophisticated medical practice in every developed country in the world.

Major improvements in transplantation for all kinds of organ transplantation were almost entirely dependent on Starzl's introduction of new drugs or drug combinations. Because of its technical simplicity, kidney transplantation continued to play a pivotal role in the clinical evaluation of these agents: adjunct antilymphocyte globulin followed by cyclosporine and tacrolimus in combination with prednisone. When cyclosporine replaced azathioprine as the baseline immunosuppression (1979–1980), and was replaced in turn by tacrolimus (1989–1990), Starzl began to use the new drugs as monotherapy, adding prednisone or antilymphocyte globulin only to treat rejection. In time, prednisone and other secondary drugs were started at the time of transplantation at most centers instead of being withheld until needed. In the United States, surgeons use the liver transplant procedure most commonly to benefit adults who have sustained liver scarring—or cirrhosis—from hepatitis C infection and children who have biliary atresia. In the United States, 6000 liver transplants are performed annually, and the number continues to rise. Liver transplant patients have approximately an 86% 1-year and a 78% 3-year survival rate.

Norman Shumway (1923–2006) the father of heart transplantation and one of the preeminent heart surgeons did not start out to become a physician. He entered the University of Michigan in 1941 intending to study law but left 2 years later after being drafted into the Army. In the military, he was given an aptitude test that asked him to check a box for a career interest: medicine or dentistry. He chose the first and was enrolled in a specialized Army program that included premedical training at Baylor University in Texas. He moved on to Vanderbilt University, where he received his MD in 1949. He did his internship and residency at the University of Minnesota, where he developed an intense interest in cardiac surgery. After another 2-year stint in the military, this time in the Air Force, he continued his surgical training in Minnesota and obtained his PhD in cardiovascular surgery in 1956. Shumway came to Stanford University in 1958 as an instructor in surgery. Shortly after his arrival, the medical school moved from San Francisco to Palo Alto, giving Shumway the opportunity to launch the cardiovascular surgery program at the new, expanded campus. In 1959, working with then surgery resident Richard Lower, he transplanted the heart of a dog into a 2-year-old mongrel. The transplanted dog lived 8 days, proving it was technically possible to maintain blood circulation in a transplant recipient and keep the donated organ alive. Shumway and his colleagues would spend the next 8 years perfecting the technique in dogs, achieving a survival rate of 60%–70%. In 1967, he announced that he was confident enough in the research to start a clinical trial and that Stanford would perform a transplant in a human patient if a suitable donor and recipient became available. But before this could take place, Christiaan Barnard of South Africa performed the world's first heart transplant on a patient who lived for 18 days, using the techniques Shumway and Lower had developed. On January 6, 1968, Shumway did his landmark human heart transplant.

Shumway and his colleagues made steady progress over the next decade through careful selection of donors and recipients, efforts to increase the donor pool, improvements in organ preservation and in heart biopsies, and advances in drugs to prevent rejection of the foreign organ, among other developments. His team was the first to introduce cyclosporine for heart transplantation in late 1980. With the availability of this immunosuppressive drug, the field took a giant leap forward. In 1981, Shumway's group performed the world's first successful combined heart–lung transplant in a 45-year-old advertising executive, Mary Gohlke, who lived 5 more years and wrote a book about her experiences. By the late 1980s, Shumway's surgical team was transplanting hearts into infants as well. In the United States, between 1988 and 2015, there were 61,595 heart transplants. Worldwide there are ~4000 heart transplants per year. The 1-year survival rates following heart transplantation have improved from 30% in the 1970s to almost 90% in the 2000s. The 5-year survival rate is ~70%.

How Does Cyclosporine Work?

Although radiation and azathioprine can act as immunosuppressives by blocking the multiplication of immune cells, they are also nonselective and can kill other proliferating body cells such as the white and red blood cells in the bone marrow and the cells lining the intestine. The oil-based formulation of cyclosporine (marketed as Sandimmune) selectively inhibits the proliferation of T lymphocytes (not B lymphocytes), thereby sparing other body cells. Although the Sandimmune formulation of the lipophilic cyclosporine contains dispersing agents that provide a certain level of oral absorption, its reliance on emulsification by bile led to patient variability in absorption. In 1995, Neoral, a microemulsion formulation of cyclosporine, was developed to provide a more consistent pattern of absorption, better pharmacokinetic predictability, and better immunosuppressive activity than Sandimmune.

Cyclosporine acts as an immunosuppressive by interfering with the production and release of the cytokine interleukin-2 through the intracellular inhibition of calcineurin; as a result, there is inhibition of lymphocyte proliferation with reduced function of T lymphocytes particularly those involved in cell-mediated (graft rejection) reactions. Cyclosporine is not lymphotoxic, and so the effects on lymphocytes are reversible.

Tacrolimus or fujimycin, first discovered in 1987 from the fermentation broth of a Japanese soil sample that contained the bacterium *Streptomyces tsukubaensis,* is also an immunosuppressive that acts in a similar fashion to cyclosporine but is much more potent. Unlike cyclosporine, it is a 23-membered macrolide lactone. The FDA first approved tacrolimus in 1994 for use in liver transplantation; this has been extended to include kidney, heart, small bowel, pancreas, lung, trachea, skin, cornea, bone marrow, and limb transplants. Immunosuppression with tacrolimus has been associated with a significantly lower rate of acute rejection compared with cyclosporine-based immunosuppression (30.7% vs. 46.4%) in one study.

The primary drawbacks of cyclosporine include renal impairment, chronic rejection, and hirsutism in children. Because of this, cyclosporine is generally used as a part of a multidrug regimen to achieve a synergistic therapy with lower toxicity. Tacrolimus has also had nephrotoxic side effects and can be diabetogenic; however, it does not cause hirsutism.

The introduction of cyclosporine was a watershed moment in the management of organ transplantation. Before its introduction, there were 10 centers in the world performing organ transplantation and within a few years there were more than 1000. In the United States, between 1988 and 2015, there were a total of 642,840 organ transplants that led to another problem: a shortage of organ donors. In addition to making the difference between life and death, cyclosporine has also enhanced our understanding of the immune system and facilitated the development of newer immunosuppressants including tacrolimus.

Detail of the painting "French psychiatrist Philippe Pinel (1745–1826) releasing lunatics from their chains at the Salpêtrière Asylum in Paris in 1795" by Tony-Robert Fleury (1837–1912). (Courtesy of the Bridgeman Art Library/Archives Charmet.)

Chapter 12

Malaria, Madness, and Chlorpromazine

Bedlam, the common name for the hospital of St. Mary of Bethlehem in London, was first founded as a religious house but was used as an asylum for the mentally ill as early as the fifteenth century. Until the middle of the twentieth century, the word "bedlam" aptly described the prevailing conditions in most psychiatric institutions where severely agitated mental patients—the insane—were detained and frequently were restrained sometimes with chains, leather straps, straitjackets, or wrapped in ice-cold sheets. By the nineteenth century, most psychiatric hospitals were for patients with paresis (dementia paralytica), a form of neurosyphilis that could develop 10–25 years after the initial infection. Patients with paresis, paretics, had the characteristics of the insane: disturbances in memory and judgment, megalomania, paranoia, violent behavior, disorientation, depression, and delusions. In the mid-twentieth century in the United States, and before there was salvarsan and penicillin to treat syphilis, general paresis of the insane (GPI) continued to be the basis for almost 10% of all the first admissions to psychiatric hospitals and accounted for more than 20% of all patients in mental hospitals.

Malariatherapy

The colonel keeps repeating "Sunshine in my head, sunshine in my head, give me a different head." He sticks out his chest, stares and shouts "Lieutenant, go tell Major Thomas Battery 155 must be transferred. They cannot fire golden shells." Everything the colonel touches turns to gold and with gold he'll buy the world. The only trouble is his head too has turned to gold so he needs sunshine, lots of sunshine to change it. The colonel's megalomaniac delirium is the result of a longstanding and neglected syphilis. He is suffering from GPI. The year is 1947, and so penicillin has not made it into general use so to make his golden head better he must be treated with fever therapy.

In 1887, Julius Wagner-Jauregg (1857–1940), a Viennese psychiatrist, addressed the problem of GPI by treating his patients with fever. Wagner-Jauregg was only the latest to use fever to treat disease. The Greek physician Hippocrates (460 BC–370 BC) observed that occasionally mental patients were healed or favorably influenced when attacked by fever. Later, Galen (130 AD–216 AD) the prominent Greek physician in the Roman Empire also found fever to have a positive effect on melancholia. During the eighteenth century, physicians following the precepts of Hippocrates and Galen tried various methods to induce fever including injections of turpentine, mercury, tuberculin, and infecting with *Salmonella typhi* to cause typhoid fever. Most of these treatments, however, failed to improve the mental condition of the patients. Then, in 1917, after nearly three decades of failed attempts Wagner-Jauregg tried malaria because of its characteristic periodic high fevers. He took blood from patients infected with *Plasmodium vivax* and injected it into the body of the GPI patient. Within a few days, the GPI patient developed a high fever lasting 5–6 h, returning to normal about 48 h later. Patients were put through this 2-day cycle 10 or 12 times with the body temperature rising to 41°C and then they were cured of malaria by being given quinine. Wagner-Jauregg reported clinical success in 6 of the first 9 patients he treated with malariatherapy. In 1927, Wagner-Jauregg received the Nobel Prize for Physiology or Medicine for "his discovery of the therapeutic value of malaria inoculation in the treatment of dementia paralytica." It was claimed that several thousand people in various clinics and asylums in Europe and elsewhere received benefits in the few years preceding the Prize; Wagner-Jauregg himself had treated over a thousand GPI cases with malaria and whereas only 1% of patients without treatment had shown full remission; with malariatherapy 30% of the cases and in some instances 50% of paretics were completely cured. The presenter at the Nobel Prize ceremony said, "It is now quite clear that Wagner-Jauregg has given us the means to a really effective treatment of a terrible disease which was hitherto regarded as resistant to all forms of treatment and incurable."

Malariatherapy was used to treat advanced neurosyphilis throughout the United States and Europe until the early 1950s, when penicillin use to cure syphilis became widespread. Whether malariatherapy was actually effective for GPI remains difficult to determine; in part, this is because it was reserved for the less seriously ill patients who were more likely to tolerate a malaria infection and neurosyphilis will vary in its severity over time and spontaneous remissions did occur. Furthermore, when malariatherapy was introduced, the concept of controlled clinical trials that might have addressed the effectiveness of the treatment had not yet been established as a standard for experimental therapeutics. Moreover, the mechanism by which fever induced by malaria might have been useful—if at all—remains to this day unclear. Malariatherapy did have some other benefits: it gave physicians and patients a sense of hope and it also encouraged a more optimistic general clinical care of patients with GPI. Perhaps the greatest benefit of malariatherapy, however, was that it could be used in the testing of putative antimalarial drugs in humans, to demonstrate strain-specific immunity to malaria, to understand the differences in therapeutic effectiveness between sporozoite and blood-induced infections, to discover the liver stages of malaria, and to isolate and identify a new human malaria, *P. ovale*.

Between 1945 and 1960—a time when malariatherapy was out of favor—there were significant changes in the therapeutic concepts and practices used in the care and management of the insane. One of the most prominent therapies was developed during the 1930s by the Viennese physician, Manfred Sakel (1900–1957). Sakel a quick tempered, arrogant psychiatrist with a dogmatic spirit wanted to shock his schizophrenic patients into normalcy. Seizing on the discovery of insulin in 1923, Sakel as an internist at the Lichterfelde Hospital for Mental Diseases in Berlin in 1927 provoked a superficial coma in a morphine-addicted woman by injecting insulin and obtained a remarkable recovery of her mental faculties. He continued the practice with patients with severe psychoses particularly schizophrenia. By 1930, he perfected what he called "Sakel's technique" to treat schizophrenics first in Vienna and later in the United States. Sakel administered large doses of insulin to his patients so that the blood glucose levels dropped precipitously. At first, the patient appeared to be asleep, then there was thirst, facial pallor, pulse and respiratory rates increased, muscles twitched, and there was profuse sweating. Then the eyes rolled, the lids closed, and there was a loss of consciousness—coma. There were no reflexes, sweating was profuse, and 1–2 h later the patient was awakened with a drink of sugar syrup. After coma, there was a profound state of regression. Sakel and other psychiatrists claimed that insulin shock therapy gave a significant improvement among schizophrenics, a group traditionally resistant to all known psychiatric interventions. And this was despite there being a mortality rate of 1%–5%! Although, as with malariatherapy, there was no scientific basis for insulin shock therapy, its alleged effectiveness promoted its widespread use.

Sakel immigrated to the United States in 1937, where he lived on Park Avenue, New York City, and treated wealthy patients in his suite and for this he received enormous honoraria. He also administered insulin coma therapy in the private Slocum Clinic in Beacon, New York, and was driven to and from his residence in a limousine provided by one of his wealthy private patients. Sakel through insulin coma therapy became quite wealthy and at his death left $2 million to his girlfriend Marianne Englander.

In Budapest, in the late 1920s, Ladislas Meduna (1896–1964) impressed by statistical studies that showed epilepsy and schizophrenia hardly ever occurred together in the same individual and schizophrenic patients who developed epilepsy seemed to be cured of their schizophrenia theorized that by inducing epilepsy he could cure schizophrenia. After undertaking preclinical trials to establish safety he stated, "I believe that this antagonism should be utilized to cure not epilepsy, but schizophrenia." In 1935, he tried injecting patients with camphor and cardiazol (pentylenetetrazol) to induce an epileptic fit. Convulsions came on quickly and violently and were dose-dependent. Meduna claimed recovery in 10 psychotic patients and no change in 12. But because the convulsions were so severe in some cases it caused spine fractures in 42% of patients.

In 1938, the newly invented electroshock apparatus, developed by Ugo Cerletti and Lucio Bini at the University of Rome, was used to induce convulsive seizures that were more reliable and less dangerous than chemically induced seizures. Electrodes were placed on the temples, a rubber gag placed in the mouth, and a small electric shock administered. The patient's face contracted in a grimace, eyes closed, jaws clenched, and muscles in the chest contracted so the patient could not breathe, then the body shuddered and shook. When the shaking subsided the patient was limp like a disjointed puppet. Breathing did not begin for 5–10 seconds and then the patient went into coma with muscular relaxation and rapid noisy breathing. After 15 minutes the patient awakened, and was hazy, fatigued, and stiff. After 10–20 electroshock treatments on alternate days, the improvements in most patients, particularly those with melancholia, was impressive.

In 1939, Meduna immigrated to the United States and took a position at Loyola University and then at the Illinois Psychiatric Institute of the University of Illinois in Chicago. Here, he began to use electroconvulsive therapy using the apparatus of Cerletti and Bini because he found it to be superior to the chemical convulsive therapy. Electroshock therapy quickly became popular among psychiatrists and the general public because it was relatively easy to administer, it appeared to be safe, and the results seemed to be promising. The fact that it caused pain and discomfort and had other undesirable effects (including broken bones) seemed to be of minor concern to the practitioners when compared to the benefits. In 1944, a committee of the American Psychiatric Association reported that electroshock therapy was effective in treating cases of severe depression, however, by 1945, the treatment had become controversial because of its promiscuous use without adjunct psychiatric therapies. Further, shock therapy provided no appreciable benefit for chronic schizophrenic patients.

Frustrated with shock therapy, some psychiatrists turned to psychosurgery. In 1935, António Egas Moniz (1874–1955), a Portuguese neurologist, performed the first operation on a patient with a cycle of obsessions. Moniz believed if he cut the path from the prefrontal lobe to the thalamus he would block the neural relay from sensory impressions to the seat of consciousness. In the procedure the patient's head was shaved and painted with iodine. Under local anesthetic skin incisions were made on the sides of the head and two small retractors were positioned to ease the way for trephination. Two holes were bored one above and in front of each ear and a trochar—a nickel-steel rod was pushed into the cerebral cortex to cut the thalamocortical fibers. For this work, Moniz received the 1949 Nobel Prize for "one of the most important discoveries ever made in psychiatric therapy." During its heyday in the 1940s and the early 1950s, prefrontal lobotomy was performed on more than 40,000 people in the United States and 10,000 in Western Europe. Lobotomy kept costs down; in the United States, the upkeep of an insane patient could cost the state ~$35,000 a year, whereas the cost of lobotomy was only $250 after which the patient could be discharged.

In some lobotomized patients there was no change, others improved, and some were reduced to a state of "placid phantom" devoid of anxiety and free of urges. They were permanently apathetic. It was claimed that lobotomized patients were more manageable and were able to adjust in ways beneficial to both the individual and the mental institution in which they were housed. But, by the late 1950s, psychosurgery was on the wane. This was not because of allegations of its brutal and offensive nature, but because after 1954, it was replaced by the use of the less invasive psychotropic drugs.

Antimalarials to Chlorpromazine

The discovery of the first family of antipsychotic drugs began in Germany in 1876 with Heinrich Caro (1834–1910) working at the BASF (Baden Aniline and Soda Factory) dye factory when he heated dimethylaminoaniline with sulfur and obtained a blue compound, later named methylene blue. Methylene blue (Figure 12.1) as we have already noted (see p. 14) was used by Paul Ehrlich in 1882 to stain malaria parasites and in 1891 he discovered the therapeutic value of methylene blue in his treatment of two sailors suffering with vivax malaria. In 1883, the chemist August H. Bernthsen (1855–1931) working with Caro at BASF reported that methylene blue was a phenothiazine derivative and he synthesized phenothiazine itself. Phenothiazine (Figure 12.1) consists of two benzene rings (=phenol) joined by a central ring containing sulfur (=thio) and a nitrogen atom (=azo). During World War I, Werner Schulemann and coworkers at Bayer synthesized derivatives of methylene blue, and one of these, the diethylaminoethyl derivative, proved to be a more active antimalarial than methylene blue; however, it was too toxic to be useful. Schulemann seized on the idea that it was the aminoalkyl side chains that were essential for antimalarial activity. He and other German chemists combined these side chains with other ring systems; in due course they found a number of antimalarials, one of which was quinacrine (marketed as Atabrine), which was used during World War II (see p. 15).

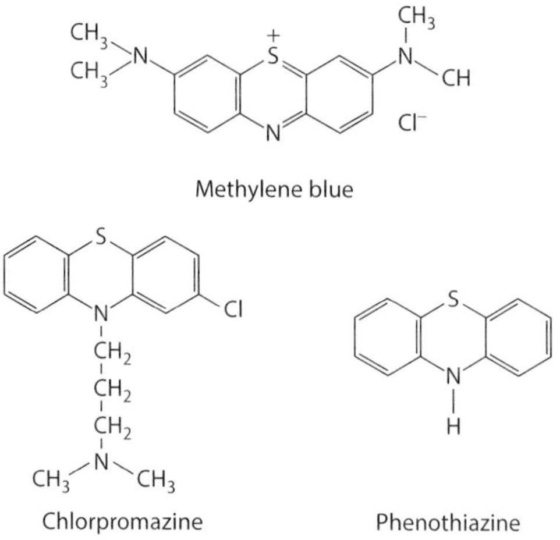

FIGURE 12.1 A comparison of the structure of methylene blue with phenothiazine and chlorpromazine.

When the Japanese occupied Java in 1942, the Allies were deprived of quinine for its military forces, and so the United States mounted a secret antimalarial program under the direction of the Office of Scientific Research and Development and the Board for the Coordination of Malarial Studies. The program's goal was to prepare and distribute putative antimalarials supplied by 20 government and nonprofit laboratories, more than 50 pharmaceutical companies, and more than 50 university and college laboratories. Supported by the program, Henry Gilman and David A. Shirley at Iowa State College began examining a series of phenothiazines to see whether they had antimalarial activity. In contrast to Schulemann, Gilman and Shirley added the amino alkyl chain not to methylene blue but to the central nitrogen atom of phenothiazine. They were disappointed to find, however, that this compound had no antimalarial activity.

Because of the disruption in international scientific communication during World War II, French chemists at Rhone-Poulenc (now Sanofi-Aventis, Paris, France), who did not learn of the work of Gilman's group on phenothiazines, continued with their research to find a suitable replacement for quinine. They confirmed Gilman and Shirley's finding that the amino alkyl derivative of phenothiazines had no effect on the symptoms of malaria; however, they elected to investigate some of their other properties particularly as antihistamines. French scientists had been interested in antihistamines since the 1930s, and in 1937, Daniel Bovet at the Pasteur Institute observed antihistamine activity for a phenothiazine compound synthesized a few years earlier by his colleague Ernest Forneau who had actually studied it for antimalarial activity. One of Forneau's compounds (F-929) was an amino alkyl substitution attached to the oxygen atom; further study had revealed that an oxygen atom could replace the nitrogen atom in the phenothiazine to produce a more potent antihistamine (F-1571). However, it was too toxic to be tested in humans. Stimulated by Bovet's results the chemists at Rhone-Poulenc prepared other compounds that were introduced as anti-allergy medications: phenbenzamine (RP-2339, Antergan), diphenhydramine (Benadryl), and a more potent antihistamine with a longer duration of action, promethazine (RP-3277, Phenergan).

On December 11, 1950, Paul Charpentier and collaborators at Rhone-Poulenc synthesized from the sedative promazine a phenothiazine with a dimethylaminopropyl side chain (RP-4560) and named it chlorpromazine (CPZ) (Figure 12.1). In May 1951, Charpentier sent CPZ to the head of pharmacology at Rhone-Poulenc, Simone Courvaisier, for animal studies. She and her group found that CPZ prolonged sleep induced by barbiturates in rodents and it had the unusual property of selectively inhibiting the conditioned-avoidance response in rats. In the conditioned-avoidance escape response test the rat was placed in a box with a grid floor through which electric shocks may be delivered. At first, the shocks were delivered at the same time a buzzer is sounded. The rat quickly learned to escape the shocks by climbing a rope suspended from the top of the box. Later, the animal became conditioned to climb the rope in response to the buzzer. CPZ inhibits the conditioned response of the rat to the buzzer, in other words with CPZ it simply ignores the sound. CPZ selectively blocks the conditioned response and is different from the action of barbiturates, which at doses sufficient to block the conditioned response also inhibits motor function so the animal is unable to climb the rope.

The potential use for CPZ in psychiatry was first recognized by Henri Laborit (1914–1995), a French naval surgeon and physiologist, during his research on artificial hibernation in the prevention of surgical shock. In 1951, Laborit in collaboration with Pierre Huguenard used CPZ as an adjunct to surgical anesthesia because of its body temperature-lowering effect, called artificial hibernation. The Laborit–Huguenard "lytic cocktail" consisted of CPZ, promethazine, and an analgesic that allowed for the use of lower doses of anesthetic agents. Laborit found that 50–100 mg of CPZ given intravenously produced disinterest in the patient without loss of consciousness and with only a slight tendency to sleep. Laborit also found CPZ to be an effective sedative that greatly reduced preoperative anxiety in surgical patients at the military hospital Val de Grace in Paris. Since cooling with water had been used in patients in France for controlling agitation, Laborit was able to persuade Joseph Hamon, Jean Paraire, and Jean Velluz at the neuropsychiatric service at Val de Grace to try CPZ in the treatment of one of their patients. On January 19, 1952, Jacques L., who was a 24-year-old severely agitated (manic) psychotic, was given 50 mg of CPZ intravenously. The calming effect was immediate but lasted only a few hours so several treatments were required. Repeated administration caused venous irritation, so on several occasions, barbiturates and electroshock therapy were used instead of CPZ. However, after 20 days of treatment with 855 mg CPZ the patient was ready "to resume a normal life." These data were reported at a meeting in February 1952 and published by Hamon and colleagues in March 1952.

Soon thereafter, the psychiatrist Pierre Deniker, Jean Delay's assistant, at Hôpital Sainte-Anne near Paris used CPZ as a monotherapy for mania and other "excited" states. The new treatment calmed the excitement and mania and diminished delusions and hallucinations. "The fury and violence had given way to calmness and peace ... the results with CPZ could be measured in the psychiatric hospital in decibels ... recorded before and after the drug. In fact, Deniker's department was a small island of silence in Sainte-Anne often filled with the cries of rage of the mentally ill patients ... ."

Delay and Deniker realized the implications of their findings and in 1952 they published a series of clinical reports of their experiences with CPZ in treating manic and psychotic patients. Their observations were subsequently confirmed by J. E. Stahelin and P. Kleinholz in Europe and Heinz Lehmann and George Hanrahan in North America. In December 1952, CPZ became available by prescription and was marketed in France by Rhone-Poulenc under the name Largactil meaning "large activity." In 1954, Smith, Kline & French (now GlaxoSmithKline) in the United States entered into an agreement with Rhone-Poulenc, and after animal testing and clinical trials involving some 2000 doctors and patients in the United States and Canada, and after FDA approval it was marketed under the trade name Thorazine for use in psychiatry and to inhibit nausea and vomiting. Within 8 months of its appearance on the market, Thorazine had been administered to over two million patients. From 1953 to 1955, CPZ treatment spread around the world. It was only in the 1960s, that the therapeutic effects of CPZ were established beyond reasonable doubt by the U.S. Veterans Administration Collaborative Study Group. However, long before that study was published large decreases in the psychiatric patient populations were seen around the world because of the widespread use of CPZ. In the United States, the number of patients in mental institutions dropped dramatically and in 1955 it was 559,000 and a decade later it was 452,000.

Aftermath

There has been confusion and dispute over who made the most significant discoveries and contributions to the use of CPZ and this may be the reason why no Nobel Prize has ever been awarded for its discovery. Despite the absence of a Nobel Prize by 1957, the importance of CPZ was recognized in the scientific community when Laborit, Lehmann, and Deniker were awarded the prestigious American Health Association's Lasker prize for their work on the clinical development of CPZ. (Laborit for using CPZ as a therapeutic agent first and then as a potential psychotropic drug, Deniker for his leading role in introducing CPZ into psychiatry and demonstrating its use in the clinical course of psychoses, and Lehmann for bringing full practical significance of CPZ to the attention of the scientific community). In 1993, other awards recognized Laborit, Hamon, Paraire, and Velluz for their role in identifying the therapeutic effects of CPZ.

Although the development of CPZ as an antipsychotic was a serendipitous finding, perhaps it should not have been a total surprise since methylene blue the original phenothiazine had been used to treat psychotic illnesses in the 1890s and tried again in the 1920s and 1930s. But it was treatment with CPZ, not methylene blue, that was the effective stimulus for academic and industrial research into other pharmacologic treatments for psychotic, mood, and other mental disorders and it opened the era of modern psychiatric chemotherapy. Although CPZ remained the most prescribed antipsychotic drug throughout the early 1960s and early 1970s, other drugs with similar antipsychotic efficacy but different chemistry and side-effect profiles were introduced and today these number in the dozens.

"CPZ made it clear that mental illness could be treated effectively by chemical means. It led to the phenomenon of deinstitutionalization of psychiatric patients and permitted many of them to be attended in their family environment and by their general physicians thus putting them on an equal footing with others, both socially and in relation to work and undoubtedly contributing to reducing the stigma associated with schizophrenia."

Epilogue

"Whoever wishes to foresee the future must consult the past"

Machiavelli

Those drugs that have changed our world will certainly guide the development of future medicines. Drugs, as we have seen, may begin as a serendipitous finding as was the case with penicillin, the anesthetics ether and chloroform, chlorpromazine, and cyclosporine. Or they may begin in a laboratory where scientists, inspired by a natural product, are able to develop a lead compound for an oral contraceptive (diosgenin to norethindrone), an antimalarial (qinghaosu to artesunate), or an analgesic (willow bark to aspirin). In some cases, the natural products themselves may be used as drugs (insulin, cocaine) or it is possible to systematically rearrange the molecular structure of a compound to produce novel antimalarials such as atabrine and chloroquine, or antibiotics such as prontosil and sulfonamides, and high-powered cocaine, heroin. Mining the specialized metabolism of HIV and to synthesize modifications of an existing molecular scaffold has produced antiretrovirals. And, with an understanding of the immune system and the characterization of antigens it has been possible to produce protective vaccines against a wide variety of pathogens.

What of the future? The discovery and development of new medicines is a long, complex, and rigorous process. Today, in the course of drug development, it is possible to screen many thousands of small molecular building blocks and other drug-like molecules (called chemical libraries) against specific biological targets—DNA, RNA, or protein enzymes. It is also possible to produce monoclonal antibodies against specific biological targets and to use recombinant DNA techniques to create novel lead compounds and to determine specific mechanisms of drug resistance.

Once a lead compound has been found, that is, a compound that kills a virus or a malaria parasite or a bacterium, then the process of optimization may begin. By rearranging the atoms in the compound (i.e., preparing analogs), a desired activity may be achieved without increasing toxicity (salvarasan to neosalvarsan) or side effects (e.g., ASA). Then too attempts can be made to improve bioavailability by changing the chemical nature of the compound itself or by altering the medium in which it is placed, for example, by formulation cyclosporine becomes Sandimmune. During these early stages of drug development, there may be laboratory and computer modeling of the target and lead compound to optimize its pharmacologic activity and by tissue culture and animal testing it may be possible to produce analogs through rational design. Pharmacokinetic studies in suitable animal models can be used to establish the absorption, distribution, metabolism, and excretion of the compound and at this stage patent protection is usually acquired. This preclinical or development phase may take 4–5 years.

If the putative drug has demonstrated the desired effects in animal tests and shows distinct advantages over existing therapeutics, has acceptable pharmacokinetics, no serious side effects and the desired half-life is achieved then it may be time for clinical trials. These trials include three phases before a drug can be approved and marketed. Phase I trials involve 100–200 healthy volunteers to establish safety, dose, and blood levels. Phase II trials involve hundreds of patient volunteers to establish drug effectiveness, short-term safety, side effects, and an optimal dose regimen. Phase III trials involve a few thousand patient volunteers to verify efficacy and monitor adverse long-term use, fine-tuning of dosage and rare side effects, and a comparison to other drugs on the market takes place during this phase. If a drug successfully passes phase III it can be registered, approved, placed on the market, and prescribed. The marketed drug will continue to be monitored for effectiveness or for rare or unexpected side effects (=phase IV). If a drug has been approved for a particular condition and is found to have more than one indication then the arsenal of drugs to combat disease is materially increased.

The development of a new drug for market may take 10–12 years and on average is said to cost a pharmaceutical company a billion dollars. It is estimated that out of every 10,000 structures synthesized, 500 will reach animal testing, 5 will be considered safe for testing in human volunteers, and only one will reach the market. In the United States, the results of all testing must be provided to the U.S. Food and Drug Administration (FDA) and/or other appropriate regulatory agencies to obtain permission (via filing of an IND, investigational new drug application) to begin testing in humans.

The Thalidomide Tragedy

The thalidomide experience is one of the darkest episodes in drug history. A West German pharmaceutical company, Chemie Grünenthal (Aachen, Germany), developed thalidomide in the 1950s to expand the company's product range beyond antibiotics. It was considered to be an anticonvulsive drug but instead it made users sleepy and relaxed. It seemed a perfect example of newly fashionable tranquilizers.

During patenting and testing, it was found that it was practically impossible to achieve a deadly overdose of the drug. Animal tests, however, did not include looking at the effects of the drug during human pregnancy. The apparently harmless thalidomide was licensed in July 1956 for prescription-free over-the-counter sale in Germany and most European countries. Because the drug also reduced morning sickness, it became popular with pregnant women. By 1960, some physicians became concerned about possible side effects. Some patients had nerve damage in their limbs after long-term use and thousands of babies worldwide were born with malformed limbs. Chemie Grünenthal did not provide convincing clinical evidence to refute the concerns. In the United States, the FDA's drug examiner did not approve the drug for use as a result American women were spared the anguish of having children being born with malformed limbs.

In Germany and elsewhere, there was an increase in births of thalidomide-impaired children. However, no link with thalidomide was made until 1961. The drug was only taken off the market in 1962, after Widukind Lenz in Germany and William McBride in Australia independently suggested the link. However, by then over 10,000 children were born with thalidomide-related disabilities worldwide.

There was a long criminal trial in Germany and a British newspaper campaign. They forced Chemie Grünenthal and its British licensee, the Distillers Company, to financially support victims of the drug. Thalidomide led to tougher testing and drug approval procedures in many countries, including the United States and the United Kingdom.

A bright side for thalidomide was discovered in 1964, when a leprosy patient at Jerusalem's Hadassah University Hospital was given thalidomide after other tranquilizers and painkillers failed. The Israeli doctor, Jacob Sheskin, noticed that the drug also reduced other leprosy symptoms. Research into thalidomide's effects on leprosy resulted in a 1967 WHO clinical trial. Positive results saw thalidomide used against leprosy in many developing countries. The positive effects are undeniable; however, the renewed use of thalidomide remains controversial. Further, there is a risk of new thalidomide births, particularly in countries where controls may not be efficient. Black market trade of unlicensed thalidomide by people with leprosy may increase the risks.

Drugs in Our Future

Natural products—from sea urchins, sea stars, sponges, marine algae, fungi, and tropical plants—will continue to serve as lead compounds. The newest therapeutics will be discovered by screening of millions of compounds in chemical libraries using robotic handling and high-throughput screens in biochemical assays. By manipulation of recombinant DNA, desired proteins at high yields will be produced. Computer simulations and modeling will also refine putative drugs and it will be possible to alter scaffolding by rational design. Site-directed mutagenesis can provide insights into mechanisms of drug resistance so that better therapeutics can be created. In addition, there will be technological breakthroughs not yet dreamed of that will lead to safe and effective therapeutics.

Although it is impossible to predict with certainty those drugs that are still beyond the horizon, it is safe to speculate that 25 years in the future, the therapeutic agents that will change the world will fall into the following categories: antibiotics with enhanced capacities to overcome resistance, antihypertensive agents, antiretrovirals, anticancer drugs, antitissue rejection drugs, immunostimulants, tranquilizers, antidiabetic drugs, and nonaddictive analgesics.

Notes

In writing this book primary and secondary literature sources and especially the *Power of Plagues* (ASM Press 2006) and *Magic Bullets to Conquer Malaria* (ASM Press 2010) have been relied upon. These chapter notes consist of references that are keyed to page numbers in the text.

Chapter 1. Malaria and Antimalarials

General

Hobhouse, H. *Seeds of Change. Six Plants That Transformed Mankind.* Counterpoint Press. Berkeley, CA, 2005.
Masterson, K. *The Malaria Project.* New American Library. New York, 2014.
Rocco, F. *Quinine: Malaria and the Quest for a Cure That Changed the World.* HarperCollins, New York, 2003.
Slater, L. B. *War and Disease. Biomedical Research on Malaria in the Twentieth Century.* Rutgers University Press, New Brunswick, NJ, 2009.

Specific

Pages 4–7. *Magic Bullets to Conquer Malaria.* pp. 23–30; *War and Disease.* Chapter 1. Quinine and the Environment of Disease. pp. 17–20.
Page 8. *Magic Bullets.* p. 159.
Pages 9–11. *Magic Bullets.* pp. 159–160.
Pages 11–20. *Magic Bullets.* pp. 52–76; *War and Disease.* Chapter 2. Avian Malaria. pp. 39–58.
Pages 21–23. *Magic Bullets.* pp. 78–80; *War and Disease.* Chapter 3. New Drugs. pp. 59–83.
Pages 23–25. *Magic Bullets.* pp. 168–177.
Pages 25–28. *Magic Bullets.* pp. 137–149.
Pages 28–29. *Magic Bullets.* pp. 168–181; The discovery of artemisinin (qinghaosu) and gifts from Chinese medicine. Youyou Tu. *Nature Medicine* 17(10):1217–1220, 2011. Semi-synthetic artemisinin, the chemical path to industrial production. J. Turconi et al. *Organic Process Research & Development* 18:417–422, 2014. K-13 propeller mutations confer artemisinin resistance to *Plasmodium falciparum*. J. Strainer et al. *Science* 347(6220):428–431, 2015. Semi-synthetic artemisinin: A model for the use of synthetic biology in pharmaceutical development. C. J. Paddon and Jay Keasling. *Nature Reviews Microbiology* 12:267–355, 2014.

Chapter 2. The Painkiller, Aspirin

General

Jeffreys, Diarmuid. *Aspirin. The Remarkable Story of a Wonder Drug.* Bloomsburg, NY.

Specific

Pages 32–33. Weissmann, G. Aspirin. *Scientific American* 264(1):84–90,1991, 2004.
Page 34. Sneader, W. The discovery of aspirin: A reappraisal. *British Medical Journal* 321:1591–1593, 2000.

Page 35. Borkin, J. *The Crime and Punishment of IG Farben*. Free Press, New York, 1978.

Page 36. Vane, J. R. Inhibition of prostglandin synthesis as a mechanism for aspirin-like drugs. *Nature* 231(25):3232–235, 1971; Vane, J. R. Back to aspirin a day? *Science* 296:474–475, 2002.

Pages 36–37. Simmons, D. L., Botting, R. M., and Hla, T. Cyclooxygenase isozymes: The prostaglandin synthesis and inhibition. *Pharmacological Reviews* 56(3):387–437, 2004.

Chapter 3. Ether, Chloroform, Cocaine, Morphine, Heroin, and Anesthesia

General

Altman, L. *Who Goes First? The Story of Self-Experimentation in Medicine*. University of California Press, Berkeley, CA, 1998.

Fenster, Julie. *Ether Day: The Strange Tale of America's Greatest Discovery and the Haunted Men Who Made It*. HarperCollins, New York, 2001.

Karch, S. B. *A Brief History of Cocaine: From Inca Monarchs to Cali Cartels: 500 Years of Cocaine Dealing*. CRC Press/Taylor & Francis, Boca Raton, FL, 2006.

Markel, H. *An Anatomy of Addiction. Sigmund Freud, William Halstead and the Miracle Drug Cocaine*. Vintage, New York, 2012.

Stratman, L. *Chloroform: The Quest for Oblivion*. Sutton, Stroud, 2003.

Streatfield, D. *Cocaine: An Unauthorized Biography*. St Martin's Press, New York, 2002.

Thatcher, V. S. *History of Anesthesia*. Garland, New York, 1984.

Wolfe, R. J. *Tarnished Idol. William Thomas Green Morton and the Introduction of Surgical Anesthesia. A Chronicle of the Ether Controversy*. Norman Publishing, San Anselmo, CA, 2001.

Specific

Pages 40–44. *The Power of Plagues*. pp. 240–244.

Page 45. Olch, P. William Halstead and Local Anesthesia. *Anesthesiology* 42(4):479–486, 1975.

Page 46. Hadda, S. Procaine: Alfred Einhorn's ideal substitute for cocaine. *The Journal of the American Dental Association* 64:841–895, 1962.

Pages 46–47. Hamblin, J. Why we took cocaine out of soda. *The Atlantic*, January 31, 2013.

Pages 48–49. Schmitz, R. Wilhelm Serturner and the discovery of morphine. *American Institute of the History of Pharmacology* 27(2):61–74, 1985.

Pages 49–50. Jenkins, P. N. Heroin addiction's fraught history. *The Atlantic* February 24, 2014; Askwith, R. How aspirin turned hero. *Sunday Times*, September 13, 1998; Scott, I. Heroin: A hundred-year habit. *History Today* 48(6):6–8, 1998.

Chapter 4. The Pill

General

Djerassi, C. *The Pill, Pigmy Chimps and Degas' Horse*. Basic Books, New York, 1992.

Djerassi, C. *This Man's Pill. Reflections on the 50th Birthday of the Pill*. Oxford University Press, New York, 2001.

Eig, J. *The Birth of the Pill*. W. W. Nortan & Company, New York, 2014.

Chapter 5. Diabetes and Insulin

General

Bliss, M. *The Discovery of Insulin.* University of Chicago Press, Chicago, IL, 2007.
Tattersall, R. *Diabetes: A Biography.* Oxford University Press, New York, 2009.

Chapter 6. Smallpox and Vaccination

General

Hopkins, D. *Smallpox.* Churchill, London, 1962.
Hopkins, D. *Princes and Peasants.* University of Chicago Press, Chicago, IL, 1983.
Preston, R. *The Demon in the Freezer.* Random House, New York, 2002.

Specific

Pages 76–85. *The Power of Plagues.* Chapter 9, Smallpox, the Spotted Plague. ASM Press, Washington, DC. pp. 191–209.

Chapter 7. Vaccines to Combat Infectious Diseases

General

Offit, P. A. *Vaccinated. One Man's Quest to Defeat the World's Deadliest Diseases.* Smithsonian Books, Washington, DC, 2008.
Oshinsky, D. *Polio: An American Story.* Oxford University Press, New York, 2005.
Smith, J. S. *Patenting the Sun: Polio and the Salk Vaccine.* Morrow, New York, 1990.

Specific

Pages 88–99. *The Power of Plagues.* pp. 12–14, 214–218, 395–399.
Page 99. Ibid., pp. 214–218.
Page 99. Ibid., pp. 395–399.
Pages 99–100. The history of the measles virus and the development and utilization of measles virus vaccines. S. L. Katz. In: *History of Vaccine Development*, S. A. Plotkin (Ed.), Springer, pp. 199–206.
Pages 100–101. The development of a live attenuated mumps virus vaccine in historic perspective and its role in the evolution of combined measles-mumps-rubella. M. Hilleman. Ibid., pp. 207–218.
Pages 101–102. History of rubella vaccines and the recent history of cell culture. S. A. Plotkin. Ibid., pp. 219–231.
Pages 102–104. The history of the pertussis vaccine. From whole cell to subunit vaccines. M. Granstrom. Ibid., pp. 73–82.
Pages 104–106. Vaccination against *varicella* and *zoster*: Its development and progress. A. Gershon. Ibid., pp. 247–264; The role of tissue culture in vaccine development. S. L. Katz, C. M. Wilfert, and F. C. Robbins. Ibid., pp. 145–149.
Pages 107–109. C. Jacobs. *Jonas Salk: A Life.* Oxford University Press. 2015.
Pages 109–112. M. Rose Jiminez. *Albert Sabin* (www.nasonline.org/memoirs).

Chapter 8. The Great Pox Syphilis and Salvarsan

Specific

Pages 116–123. *The Power of Plagues.* Chapter 12. The Great Pox Syphilis. pp. 255–274.

Chapter 9. Prontosil, Pyrimethamine, and Penicillin

General

Clark, R. W. *The Life of Ernst Chain: Penicillin and Beyond.* St. Martins Press, New York, 1985.
Greenwood, D. *Antimicrobial Drugs: Chronicle of a Twentieth Century Medical Triumph.* Oxford University Press, New York, 2008.
Hager, T. *The Demon under the Microscope.* Harmony Books, New York, 2006.
Hare, Ronald. *Birth of Penicillin.* George Allen & Unwin, London, UK, 1970.
Lax, Eric. *The Mold in Dr. Florey's Coat.* Henry Holt, New York, 2005.
McFarlane, R. G. *Alexander Fleming: The Man and the Myth.* Harvard University Press, Cambridge, MA, 1984.
Macfarlane, R. G. *Howard Florey: The Making of a Great Scientist.* Oxford University Press, Oxford, 1979.

Specific

Pages 126–133. Magic Bullets to Conquer Malaria. pp. 81–93.
Pages 133–136. Ibid., pp. 94–100.
Pages 139–144. Lax, Eric. *The Mold in Dr. Florey's Coat*; McFarlane, R. G. *Alexander Fleming: The Man and the Myth.*

Chapter 10. AIDS, HIV, and Antiretrovirals

Specific

Pages 148–150. *The Power of Plagues.* Chapter 5. A Modern Plague, AIDS. pp. 89–116.
Pages 150–151. Mohammedi, P. et al. 24 hours in the life of HIV-1 in a T-cell line. *PLoS Pathogens* 9(1):e1000361, 2013.
Pages 151–152. Molotsky, I. U.S. approves drug to prolong lives of AIDS patients. *New York Times* March 21, 1987.
Pages 153–154. Avery, Mary Ellen. Gertrude Elion Biographical Memoirs National Academy of Science; Elion, G. The purine path to chemotherapy. Nobel lecture. December 8, 1988.
Pages 154–155. Arts, E. J. and Hazuda, D. J. HIV-1 Antiretroviral drug therapy. *Cold Spring Harbor Perspectives in Medicine* 2(4):a007161, 2012.

Chapter 11. Organ Transplantation and Cyclosporine

Specific

Pages 160–161. Calne. R. Y. It can't be done. *Nature Medicine* 18(10):xxiii–xxv, 2012.
Pages 161–165. Borel, J. and Kis, Z. L. The discovery and development of cyclosporine (Sandimmune). *Transplantation Proceedings* 23(2):1867–1874, 1991; Heusler, K. and Pletscher, A. The controversial early history of cyclosporine. *Swiss Medical Weekly* 131:299–302, 2001; Borel, J. et al. The history of the discovery and development of cyclopsporine (Sandimmune). In: Merluze, V. and Adams, J. (Ed). *The Search for Anti-Inflammatory Drugs,* Birkhauser, Boston, MA, 1995. pp. 27–64.

Page 163. Kolata, G. Drug transforms transplant medicine. *Science* 221:340–342, 1983.

Page 164. Borel, J. F. et al. In vivo pharmacological effects of cyclosporine and some analogues. *Advances in Pharmacology* 35:115–246, 1996.

Page 164. Stahelin, H. The history of cyclosporine A (Sandimmune) revisited: Another point of view. *Experientia* 52:5–13, 1996.

Page 165. Kyriakides, H. G. and Miller, J. Use of cyclosporine in renal transplantation. *Transplantation Proceedings* 36(Suppl. 23):167S–172S, 2004.

Pages 165–166. Starzl, T. E. History of clinical transplantation. *World Journal of Surgery* 24(7):759–782, 2000.

Pages 166–167. Fitzpatrick, L. Heart transplants. *Time,* November 16, 2009; Norman Shumway. Heart transplantation pioneer dies at 83 (Stanford Medicine News Center. February10, 2007).

Pages 167–168. Zirkle, C. L. To tranquilizers and anti-depressants from antimalarials and anti-histamines. In: *How Modern Medicines Are Discovered,* Clarke, F. E. (Ed). Futura, Mt. Kisco, NY, 1973.

Chapter 12. Malaria, Madness, and Chlorpromazine

General

Ban, T. A. Fifty years of chlorpromazine: A historical perspective. *Neuropsychiatr. Dis. Treat.* 3(4):495–500, 2007.

Frankenburg, F. and Baldessarini, R. J. Neurosyphilis, malaria and the discovery of antipsychotic agents. *Harv. Rev. Psychiatry* 16:299–307, 2008.

Shen, W. A history of anti-psychotic drug development. *Compr. Psychiatry* 40(6):407–414, 1999.

Specific

Page 172. Thuiller, J. *Ten Years That Changed the Face of Mental Illness.* Blackwell, New York, 1999.

Page 173. Chernin, E. The malariatherapy of neurosyphilis. *J. Parasitol.* 70(50):611–617, 1984.

Page 173. Shute, P. G. Thirty years of malaria-therapy. *J. Trop. Med. Hyg.* 61:57–61, 1958.

Page 173. Wernstadt, W. Presentation speech. Nobel lecture Physiology or Medicine. December 10, 1927. Elsevier, Amsterdam, the Netherlands.

Pages 173–174. Fink, M. Meduna and the origins of convulsive therapy. *Am. J. Psychiatry* 141(9):1034–1041, 1984; Shorter, E. Sakel versus meduna. *JECT* 25(1):12–14, 2009.

Pages 174–175. Grob, G. N. *From Asylum to Community.* Princeton University Press, Princeton, NJ, 1991. pp. 124–156; Wright, B. A. An historical review of electroconvulsive therapy. *Jefferson J. Psychiatry* 11:68–74, 2011.

Pages 175–179. Lopez-Munoz, F. et al. History of the discovery and clinical introduction of chlorpromazine. *Ann. Clin. Psychiatry* 17(3):113–135, 205; Lehmann, H. and Ban, T. The history of psychopharmacology of schizophrenia. *Can. J. Psychiatry* 42:152–162, 1997.

Epilogue

Brown, E. D. and Wright, G. D. Antibacterial drug discovery in the resistance era. *Nature* 529:336343, 2016.

Snape, T. J. and Astles, A. M. The process of drug development from laboratory bench to the market. *Pharmaceutical Journal* 285:272, 2010.

Index

Note: Page numbers followed by f and t refer to figures and tables, respectively.